*Chaotic evolution and
strange attractors*

Lezioni Lincee

Editor: Luigi A. Radicati di Brozolo, Scuola Normale Superiore, Pisa

This series of books arises from a series of lectures given under the auspices of the Accademia Nazionale dei Lincei through a grant from IBM.

The lectures, given by international authorities, will range on scientific topics from mathematics and physics through to biology and economics. The books are intended for a broad audience of graduate students and faculty members, and are meant to provide a *'mise au point'* for the subject they deal with.

The symbol of the Accademia, the lynx, is noted for its sharp sightedness; the volumes in the series will be penetrating studies of scientific topics of contemporary interest.

Chaotic evolution and strange attractors

The statistical analysis of time series for deterministic nonlinear systems

DAVID RUELLE

Professor at the Institut des Haute Edudes Scientifiques, Bures-sur-Yvette

Notes prepared by Stefano Isola
from the 'Lezioni Lincee' (Rome, May 1987)

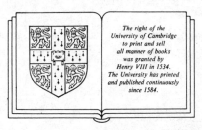

The right of the
University of Cambridge
to print and sell
all manner of books
was granted by
Henry VIII in 1534.
The University has printed
and published continuously
since 1584.

CAMBRIDGE UNIVERSITY PRESS

Cambridge

New York   Port Chester

Melbourne   Sydney

Published by the Press Syndicate of the University of Cambridge
The Pitt Building, Trumpington Street, Cambridge CB2 1RP
40 West 20th Street, New York, NY 10011 USA
10 Stamford Road, Oakleigh, Melbourne 3166, Australia

First published 1989

Printed in Great Britain at the University Press, Cambridge

British Library cataloguing in publication data

Ruelle, David
Chaotic evolution and strange attractors
the statistical analysis of time series
for deterministic nonlinear systems.
1. Nonlinear dynamical systems. Chaotic behaviour.
I. Title   II. Isola, S. (Stefano).
III. Accademia nazionale dei Zincei.
IVS Series
515.3′5

Library of Congress cataloguing in publication data

Ruelle, David.
  Chaotic evolution and strange attractors : the statistical
analysis of time series for deterministic nonlinear systems / David
Ruelle ; notes prepared by Stefano Isola from the Lezioni Lincee,
Rome, May 1987.
      p.   cm.
  Collects together a series of lectures given by David Ruelle at
the Accademia dei Lincee (Rome, May 1987) – Foreword.
  Bibliography: p.
  Includes index.
  ISBN 0 521 36272 5.   ISBN 0 521 36830 8 (paperback)
  1. Differentiable dynamical systems.   2. Ergodic theory.
3. Chaotic behavior in systems.   I. Title.
QA614.8.R83 1989   88-20317
515.3′5–dc 19   CIP

ISBN 0 521 36272 5 hard covers
ISBN 0 521 36830 8 paperback

CONTENTS

FOREWORD

The theory of differentiable dynamical systems is a rich and diversified field, emerging somewhere between mathematics and the sciences, where it constitutes an important conceptual scheme for the 'decodification' of many dynamical processes in a unified way. Indeed, an attentive analysis shows that many different systems, coming from physics, biology, chemistry, economics and technology, have time evolutions that, although irregular and chaotic, nevertheless make them similar under many qualitative aspects. However, this similarity does not appear in a straightforward manner. Rather, it emerges within a conceptual frame where many mathematical ideas and techniques, pertaining number theory, topology, probability theory, etc. are brought together to form a complex and fascinating theoretical *corpus*.

This review, which collects together a series of lectures given by David Ruelle at the Accademia dei Lincee (Rome, May, 1987), deals with those aspects of dynamical systems which are more closely related with ergodic theory, namely with the study of the properties of the invariant measures generated by the time evolutions themselves.

In writing this work, I strove to make it as structured and progressive as possible in order to make it accessible to readers not yet quite at their ease with these topics.

With this in mind I divided the text into two parts. The first part attempts to clarify the interpretative frame in which several dynamical processes (natural as well as 'artificial') can be approached by means of mathematical models of a deterministic type. In the second part the concept of invariant probability measure is introduced, along with some ergodic quantities such as characteristic exponents, entropy, dimensions, resonances, etc., which make it possible to extract useful information on the asymptotic statistical properties of the time evolutions we are dealing with.

Furthermore, I included a number of examples and figures in such a way as to give some references, useful both as clarifying elements and to emphasise the deep mathematical ideas which permeate the theory of differentiable dynamical systems.

I wish to thank Giovanni Gallavotti and Luigi Arialdo Radicati who made it possible for me to carry out this work. My special thanks go to David Ruelle for his precious advice and remarks.

Stefano Isola
Florence, December 16, 1987

INTRODUCTION

In recent years, actually the last two decades, an increasing amount of interest has been addressed to the fact that, in many different domains of science, systems with a similar strange behavior are frequently encountered. These systems display a so-called *chaotic* time evolution. Indeed, the irregular behavior of a larger and larger class of phenomena can be described with a relatively small class of mathematical objects, each of them specifying an iteration procedure that, despite its deterministic nature, produces time evolutions which are really unpredictable. Let us make this more precise by saying that the nonlinearities present in the problem can produce an extreme sensitivity with respect to the initial state of the system, so that even a very small error in its knowledge will be exponentially amplified as the time goes on. Then, a fundamental question is whether it is possible to extract useful information on the motion from experimental observations.

The present review will lead us through a progressive discussion of those theoretical and experimental aspects which constitute, at present time, the main and successful ideas on the *statistical* analysis of time series for deterministic nonlinear systems. By *time series* we mean sequences of data representing the time evolution of one or more observables of the system, monitored at fixed time intervals. On one hand, there are no constraints on which kinds of phenomena this time evolution is representing – it may describe situations emerging from experimental physics, economics, biology, chemistry, and so on; on the other hand, the claim is that its qualitative features are the same as those of a differentiable dynamical system. In other words, the starting point in this picture is that, given some initial data, an evolution equation of the form

$$\frac{d\mathbf{x}(t)}{dt} = F(\mathbf{x}(t)) \text{ with } \mathbf{x} \in R^m \tag{1}$$

in a space of suitable (small) dimension m (it might be a compact manifold M), can *model* the behavior one deals with, as soon as one identifies the vector $\mathbf{x}$ as the set of observables describing the state of the system, and the differentiable function F determining the nonlinear time evolution of these observables. More generally we call *observable* any differentiable function $g: R^m \to R$. Then, an experiment consists in picking an initial point $\mathbf{x}(0) \in R^m$ (or M) and plotting $G(t) = g(\mathbf{x}(t))$ as a function of t. The challenge is to extract physical reality from such a graph.

Now, experiments with dynamical systems usually exhibit transient behavior followed by an asymptotic motion lying on an *attractor* in phase space, namely a subset of R^m on which the orbits converge, covering it densely as the time goes on. One can provide a classification of the different levels of complexity of motions by looking at the *geometric* structure of the attractors; for example, one may see fixed points, limit cycles (periodic attractors), quasiperiodic attractors or strange attractors. However, there are many situations in which such a geometric description is no longer feasible, due either to the extreme complication of the dynamics or to the relatively high dimensionality of the problem. These cases are those in which statistical analysis becomes really relevant. As a matter of fact, the statistical theory of dynamical systems also investigates those properties of the motion which hold asymptotically, but it focuses its attention on *invariant measures* rather than on attractors. Generally speaking, if we analyse a sufficiently long record of a chaotic signal generated by a deterministic time evolution, we may find that the signal amplitudes are within definite ranges for a well defined fraction of the time; and analysis of other records of this signal may yield the same fraction. More precisely, this means that the experimental time average, i.e. a limit of the form

$$\lim_{T \to \infty} \frac{1}{T} \int_0^T g(\mathbf{x}(t)) \mathrm{d}t$$

does exist for suitable initial points in R^m, producing a well defined invariant probability measure ρ (in the sense that the time average of an observable g is equal to its space average $\rho(g)$).

The existence of such a *physical measure* ρ makes us able to use *ergodic theory* as a tool to discuss those invariant statistical properties of the time evolution like entropy, characteristic exponents, information dimension and so on.

This review is structured as follows: In part I we will get closer to the concept of chaos for deterministic systems through a survey of some 'historical' topics, like, for example, the interpretation of hydrodynamical turbulence. Some definitions of 'geometrical' notions related to chaotic phenomena, like those of strange attractor, fractal dimensions, reconstruction of the dynamics from a time series and so on, are also given.

In part II we will introduce the notion of invariant probability measure and those quantities which are, at present, the relevant tools of the ergodic theory of chaos: characteristic exponents, entropy, information dimension, resonances, and related quantities.

For more details on many arguments discussed here, we refer to the review 'Ergodic theory of chaos and strange attractors' by J.-P. Eckmann and D. Ruelle (1985).

Part I Steps to a deterministic interpretation of chaotic signals

1

Descriptions of turbulence

An important question arising from many experimental situations (for example, studying turbulent behavior in a fluid flow) is the following: How does one explain a situation in which one gets a signal (i.e. a time series) which is nonperiodic, indeed a chaotic signal?

One idea is the following: if one has a system which gives a 'noisy' signal, that is nonperiodic and irregular, this means that there must be some inputs which are also noisy and nonperiodic. The formalisation of this idea lies on the so-called *stochastic evolution equations*, namely equations of the form

$$\frac{d\mathbf{x}(t)}{dt} = F(\mathbf{x}(t)) + w(t), \tag{2}$$

where $w(t)$ is the noise term (e.g. a stochastic process).

As far as hydrodynamical systems are concerned, the (infinite) set of observables $\mathbf{x}(t)$ will represent the modes amplitude (to be defined later), and an experiment carried out with a sufficiently excited system (for example high values of the Rayleigh number) yields a situation just like the one sketched in Fig. 1. Let us stress that turbulence is the

Fig. 1. A chaotic signal obtained by the simple deterministic difference equation $x_{n+1} = 4x(1 - x_n)$. Any correlation test would reveal rapid decorrelation between successive iterations, making this sequence akin to a random sequence.

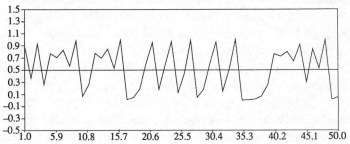

type of physical phenomena where one gets such a noisy signal. It is interesting to note that some people have considered that the theory of turbulence must necessarily be of the form expressed by (2).

Nevertheless, there are other kinds of explanations of a non-periodic signal. One of them involves the presence of many oscillators (Landau). According to this theory, the time evolution of the physical parameter describing a turbulent fluid is given (asymptotically) by:

$$\mathbf{x}(t) = \mathbf{f}(\omega_1 t, \omega_2 t, \ldots, \omega_k t) = \mathbf{f}(\phi_1, \phi_2, \ldots, \phi_k), \tag{3}$$

where $\mathbf{f}$ is a periodic function of period 2π in each of its arguments, and $\omega_1, \omega_2, \ldots, \omega_k$ are rationally independent frequencies; $\mathbf{x}(t)$ is then a *quasiperiodic* function of t. The motion (3) describes a k-dimensional torus T^k (i.e. the product of k circles) embedded in R^m and constitutes what is called a *quasiperiodic attractor*. Such attractors are a generalisation of periodic orbits, but they describe motions which look indeed nonperiodic and very irregular, thus suggesting turbulence.

In general, in a quasiperiodic time evolution it does not make sense to specify which are *the* frequencies $\omega_1, \ldots, \omega_k$ of the motion. In fact, if we look at the Fourier transform of the signal $\mathbf{x}(t)$:

$$\mathbf{x}(t) = \sum_{n_1 \ldots n_k} \hat{x}_{n_1 \ldots n_k} \exp[i(\omega_1 n_1 + \cdots + \omega_k n_k)t] \tag{4}$$

we can see that all the harmonics are present. Therefore, we can choose any other set of 'basic' frequencies of the form:

$$\omega'_j = n_{j1}\omega_1 + \cdots + n_{jk}\omega_k \quad \text{with } j = 1 \ldots k, \tag{5}$$

where the matrix (n_{ij}) has integer entries and determinant ± 1. Then, the *number* k of frequencies of a quasiperiodic motion is defined as the minimum number of rationally independent frequencies of the form (5) which are present in the Fourier transform (4). This is just what is referred to as the number of *modes* of the system, and, in a sense, it plays the role of effective 'dimension' of a quasiperiodic motion.

We shall see later that, as far as nonlinear dynamical systems are concerned, even a finite-dimensional motion need not be quasi-periodic (indeed it may be chaotic), and the concept of 'number of modes' must be replaced by other concepts such as 'information dimension', or 'number of non-negative characteristic exponents'.

Now, starting from the evolution $\mathbf{x}(t)$ of coordinates, we can introduce a more general *observable*:

$$G(t) = g(\mathbf{x}(t)), \tag{6}$$

where $g: R^m \to R$ is a differentiable function. Then, a first indicator of the qualitative nature of the motion is the *power spectrum* or *frequency spectrum* (see Fig. 2), which measures the amount of energy per unit time contained in the signal $g(t)$ as a function of the frequency ω:

$$S(\omega) = (\text{const.}) \lim_{T \to \infty} \frac{1}{T} \left| \int_0^T G(t) \exp(-i\omega t) dt \right|^2. \tag{7}$$

Fig. 2. (a) The spectra of the velocity of a fluid occupying the interval between two coaxial cylinders, when the inner cylinder is rotated at three different speeds (Couette flow). (b) The spectra of the convective heat transport in a liquid layer heated from below at three different heating intensities (Rayleigh–Bénard convection). In both cases, from the top to the bottom we see a periodic spectrum, a quasiperiodic spectrum and a continuous spectrum. From Gollub and Swinney (1978).

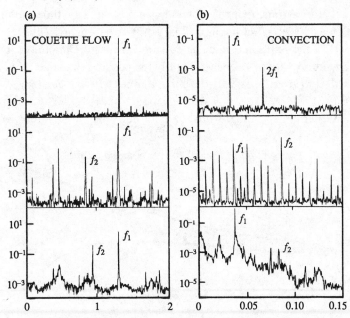

In the particular case of quasiperiodic time evolutions we find that $S(\omega)$ is formed of discrete peaks corresponding to the basic frequencies $\omega_1 \ldots \omega_k$ and their linear combination with integer coefficients

$$S(\omega)= \sum_{n_1 \ldots n_k} c_{n_1 \ldots n_k} \delta\left(\omega - \sum_{i=1}^{k} n_i\omega_i\right). \tag{8}$$

However, in practice one never computes the limit in (7), therefore the peaks have, at least, a width $2\pi/T$. Moreover, the structure of the spectrum, namely the number of peaks which are really visible, changes with the choice of the function g, and experimentally it is difficult to observe more than a few independent frequencies.

Now, the main problem with a quasiperiodic theory of turbulence (putting several oscillators together) is the following: when there is a nonlinear coupling between the oscillators, it very often happens that the time evolution does not remain quasiperiodic. As a matter of fact, in this latter situation, one can observe the appearance of a feature which makes the motion completely different from a quasiperiodic one. This feature is called *sensitive dependence on initial conditions* and turns out to be the conceptual key to reformulating the problem of turbulence.

Let us assume that the system has a deterministic time evolution defined by an autonomous ordinary differential equation like (1). Let $\mathbf{x}(t)$ be the solution of such an equation corresponding to the initial condition $\mathbf{x}(0)$. If we change slightly the position of the initial point: $\mathbf{x}(0) \rightarrow \mathbf{x}'(0) = \mathbf{x}(0) + \delta\mathbf{x}(0)$, the point at time t will also be changed (see Fig. 3).

Generally speaking, from the continuity of the solutions of an

Fig. 3. Effect of a small change of initial condition.

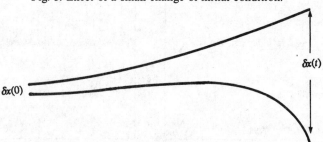

ordinary differential equation with respect to the initial conditions, one expects that, if $\delta\mathbf{x}(0)$ is small, $\delta\mathbf{x}(t)$ is also small. But, what may happen is that, when the time becomes large, the small initial distance grows anyway, and it may grow exponentially fast: $\delta\mathbf{x}(t) \sim \delta\mathbf{x}(0)\exp(\lambda t)$, where λ measures the mean rate of divergence of the orbits. In this case the motion, although purely deterministic, has those stochastic features referred to as *chaos*. In fact, in all those cases in which the initial state is given with limited precision (if we assume that the space–time is continuous this is always the case because a generic point turns out to be completely specified only by an infinite amount of information, for example by an infinite string of numbers), we can observe a situation in which, when time becomes large, two trajectories emerge from the 'same' initial point. So, even though there is a deterministic situation from a mathematical point of view (the uniqueness theorem for ordinary differential equations is not in question), nevertheless the exponential growth of errors makes the time evolution self-independent from its past history and then nondeterministic in any practical sense.

However, it is quite obvious that a quasiperiodic motion represented by a solution like (3) cannot exhibit sensitive dependence on initial conditions. A small change in initial conditions simply replaces the arguments $\omega_1 t, \ldots, \omega_k t$ by $\omega_1 t + \alpha_1, \ldots, \omega_k t + \alpha_k$, where $\alpha_1, \ldots, \alpha_k$ are small.

We shall see that a theoretical approach which wants to describe in a coherent fashion some hydrodynamic phenomena, in particular which wants to explain a noisy signal like the one in Fig. 1, has to put itself in the picture of deterministic noise; so that the meaning of turbulence will get close to those of chaos, dynamical instability and strange attractors.

The existence of sensitive dependence on initial conditions was first noticed by Hadamard, at the end of the last century, when studying the geodesic flow on compact surfaces of constant negative curvature. Such a compact surface M is obtained from the Lobachevsky plane by making certain identifications. The Lobachevsky plane itself may be viewed as the complex upper half-plane $\mathrm{Im}\,z > 0$ with the metric $\mathrm{d}s^2 = (\mathrm{d}x^2 + \mathrm{d}y^2)/y^2$, where $z = x + \mathrm{i}y$. The geodesics are then the half-circles and straight lines orthogonal to the x-axis. Returning to the compact manifold M, we let $T_1 M$ be the set of vectors tangent to M and of length 1. An element (x,u) of $T_1 M$ is thus a vector tangent to M at some point x. There is a unique oriented geodesic of M passing through x and tangent to u. Let x_t be the point at distance t of x on the geodesic, and u_t the unit vector tangent to the geodesic at x_t. The map which sends (x,u) to (x_t, u_t) is a diffeomorphism f^t of $M_1 T$ (differentiable map with differentiable inverse), and the family (f^t) is the geodesic flow. This is a particular case of an Anosov flow on $T_1 M$, which means that it has the following remarkable property: the tangent spaces to $T_1 M$ can be written as direct sums $E^s + E^u + E^0$ where E^0 is one-dimensional in the direction of the flow and E^u (respectively E^s) is exponentially expanded (respectively contracted) by the flow. We shall come back later to a more precise discussion of these properties but, for the moment, we can get an intuitive idea of the dynamical instability of a hyperbolic flow by looking at Fig. 4.

After Hadamard had realised the possible presence of dynamical instability due to sensitive dependence on initial conditions, Poincaré and Duhem wrote popular texts explaining the philosophical importance of this feature. Then, although this was not forgotten in mathematics, it seems that for a long period it was forgotten by physicists.

The rediscovery of sensitive dependence on initial conditions, about 25 years ago, corresponded to the availability of electronic computers, which allowed the 'step-by-step' computation of the solutions of differential equations. This kind of computation has shown, and continues to show with strong evidence, that many time

Fig. 4. The t-axis (the direction of the flow) is the intersection of two surfaces of trajectories approaching it as $t \to \infty$ (the (x,t) surface) and as $t \to -\infty$ (the (y,t) surface); the remaining trajectories move away both for $t \to \infty$ and $t \to -\infty$. From Arnold (1980).

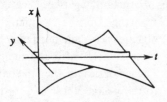

evolutions related to physical as well as technological problems do exhibit chaotic behavior, i.e. sensitive dependence on initial conditions.

A celebrated example is the Lorenz system (see Lorenz, 1963), a nonlinear time evolution in R^3 defined by the equations:

$$\frac{dx}{dt} = -\sigma x + \sigma y$$

$$\frac{dy}{dt} = -xy + rx - y \tag{9}$$

$$\frac{dz}{dt} = xy - bz.$$

These equations are obtained by truncation of the Navier–Stokes equation, and give an approximate description of a horizontal fluid layer heated from below. The warmer fluid formed at the bottom tends to rise creating convection currents. This is similar to what happens in the earth's atmosphere. For sufficiently intense heating the time evolution has sensitive dependence on initial conditions, thus representing a very irregular and chaotic convection. This fact was used by Lorenz to justify the so-called 'butterfly effect', a metaphor of the imprecision of weather forecasting. Actually, how this system really relates to turbulence (in particular to atmospheric turbulence) is not known yet, but what is remarkable is that it gives rise to a type of attractor which is nonclassical (neither periodic nor quasiperiodic).

The geometrical object described by the points of the trajectory is our first example of a *strange attractor*. We shall worry later about the mathematical definition of a strange attractor; for the moment, let us say that it is an infinite set of points (of which Fig. 5 shows a subset), in an m-dimensional space (here $m = 3$), which represents the asymptotic behavior of a chaotic system.

It is worth remarking that a time evolution which is chaotic in the sense sketched above usually exhibits a continuous power spectrum. On the other hand, the power spectrum of the velocity of a turbulent fluid is found experimentally to be continuous (see, for example, the bottom of Fig. 2). At first, this fact was attributed to the accumulation of a large number of independent frequencies, but accurate experiments by Swinney have shown that, when the fluid is excited above a certain threshold, a sharp transition towards a really continuous spectrum takes place.

So far, the power spectrum is the first indicator we have introduced which enables us to *distinguish* regular and chaotic motions. However, it is not really a 'good' indicator for the specific analysis of chaotic motions on strange attractors because the 'dimension' of a chaotic motion is no longer related to the number of independent frequencies (i.e. the number of modes); rather, it constitutes an important statistical feature in itself, which can be related both to the temporal aspect of chaos (number of positive characteristic exponents) and to its geometrical aspect (scale laws characterising the self-similar structure of strange attractors).

Fig. 5. The trajectory of the points (x,y,z) corresponding to the solutions of (9) with initial conditions near $(0,0,0)$ and with $\sigma = 10$, $b = 8/3$, $r = 28$. From Lanford (1977).

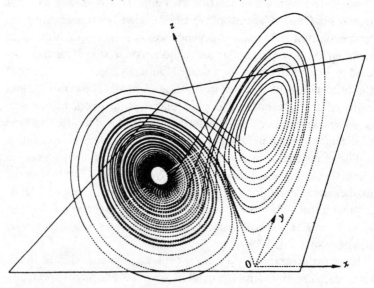

2

A bit more on turbulence. The Navier–Stokes equation

Let us give an overview of the general mathematical formulation of turbulence in such a way as to make more precise the nature of the time evolution of a hydrodynamic system.

Consider a viscous incompressible fluid occupying a region $\Omega \subset R^d$, where d may be 2 or 3. If thermal effects are ignored, this system is described by its velocity at every point of Ω, i.e. by a velocity field $\mathbf{v}$ over Ω. The time evolution of each component v_i, $i = 1 \ldots d$, of such a field is described by a partial differential equation, which is called the Navier–Stokes equation:

$$\frac{\partial v_i}{\partial t} + \sum_j v_j \frac{\partial v_i}{\partial x_j} = v \Delta v_i - \frac{1}{\rho} \partial_i p + f_i, \tag{10}$$

together with the incompressibility condition:

$$\sum_{i=1}^{d} \frac{\partial v_i}{\partial x_i} = 0, \tag{11}$$

where $\mathbf{f}$ is the external force per unit volume, ρ is the constant density, p is the pressure and v is the constant kinematic viscosity. Note that if $v = 0$ in (10) one gets the Euler equation. The Navier–Stokes equation has been derived physically as a first order approximation. In fact, looking at (10) we see that the acceleration $d\mathbf{v}/dt$ is given by an external force term, a pressure term and a friction term which is computed invoking *linear* response theory. This makes (10) a phenomenological equation, and therefore one has to investigate its mathematical properties in order to consider those situations which can be really described by it.

Of particular physical interest is the initial boundary value problem on Ω:

$$\begin{aligned} &\mathbf{v}(x,0) = \mathbf{v}_0(x) \text{ if } x \in \Omega \\ &\mathbf{v}(x,t) = 0 \text{ if } x \in \partial\Omega,\, 0 < t < T. \end{aligned} \tag{12}$$

Here the initial data should satisfy

$$\mathbf{v}_0 = 0 \text{ on } \partial\Omega, \text{ and } \partial\cdot\mathbf{v}_0 = 0. \tag{13}$$

In order to make this problem easier, one can decompose the (infinite-dimensional) space $V(\Omega)$ of the vector fields $\mathbf{v}$ over Ω as the direct sum of two subspaces: $V = H \oplus K$; here H is the subspace of the fields $\mathbf{v}$ with divergence zero, while $\mathbf{v} \in K$ if $\mathbf{v} = \partial g$, K is the subspace of the velocity fields which are gradients. Projecting (10) in the subspace H one gets an equation without the pressure p, which is the equation often actually used for the study of hydrodynamics in two and three dimensions. The main question arising at this point concerns existence and uniqueness of solutions for the initial boundary value problem (12). The answer to this question is found to depend strongly on the dimensionality d: it is a long time that we have a good existence and uniqueness theorem for the case $d = 2$ (Leray, Ladyzhenskaya), but for $d = 3$ we do not have a general theorem asserting simultaneously existence and uniqueness. More precisely, even though the existence of *weak* solutions was proved (Leray, Hopf), the theory with $d = 3$ remains fundamentally incomplete because it is not known whether or not the velocity $\mathbf{v}$ develops singularities (becoming infinite) after a finite time, even with 'good' initial data: $\mathbf{v}_0 \in C^\infty(\Omega)$. If singularities occur, the solution may be nonunique. Therefore, the problem arises to evaluate the times at which singularities occur, and Leray found that they are a relatively small subset of all times. After that, nothing much better was done for a long period, but recently V. Scheffer obtained some stronger results studying not only the times at which singularities occur but their location in space–time. A point (x,t) is called *singular* if $\mathbf{v}$ is not L^∞_{loc} in any neighborhood of (x,t); otherwise (x,t) is called a *regular* point. Scheffer's result was a partial regularity theorem, namely an estimate for the dimension of the set S of singular points. Somewhat later, using similar ideas other authors (see Caffarelli, Kohn and Nirenberg, 1982) obtained the result which appears to be the best at present.

Let us introduce a notion which will be useful in the following.

Given a nonempty set S, with a metric, and $r > 0$, denote by σ a covering of S by a countable family of subsets σ_k of diameter $r_k \leq r$. Given a real number $\alpha \geq 0$, we define:

$$m^{\alpha}(S) = \lim_{r \to 0_+} m_r^{\alpha}(S), \tag{14}$$

where

$$m_r^{\alpha} = \inf_{\sigma} \left\{ \sum_{k=1}^{\infty} (r_k)^{\alpha} \right\}. \tag{15}$$

It is clear that the limit in (14) exists – finite or infinite; in fact, when r decreases, the infimum in (15) extends over smaller and smaller classes of coverings and hence m_r^{α} increases – or at least does not decrease. The number m^{α} is called the *Hausdorff measure in dimension* α. Then, for any weak solution of the Navier–Stokes equation with $d = 3$, the singular set $S \subset R^3 \times R$ (with the Euclidean metric) has one-dimensional Hausdorff measure zero; roughly speaking it represents less than a line in four-dimensional space–time.

At present, further studies on this problem are in progress, in particular by Scheffer. In general the situation gets worse as the dimension increases, and this makes the problem of existence and uniqueness for the solutions of the Navier–Stokes equation in three dimensions a real challenge from a mathematical point of view. Nevertheless, the whole problem may be nonphysical. As a matter of fact, when one looks at a real fluid, completely different problems arise in order to explain its behavior. For example, when studying a real fluid, one has to avoid large negative pressure inside it, in order to avoid the formation of bubbles (cavitation). This problem is completely ignored in the context of mathematical discussions of existence and uniqueness of solutions. Moreover, it seems that the so-called weak turbulence, which is extensively studied experimentally, is far from the situation in which solutions develop singularities. About strong, or fully developed, turbulence, not much can be said at this time.

Although some authors continue to associate turbulence with the presence of singularities in the solutions of the Navier–Stokes equation, it should be pointed out that weak turbulence, i.e. chaotic behavior of fluid flows, has not very much to do with the specific mathematical problems given by the Navier–Stokes equation, and it can be described by simpler and handier equations of the form

$$\frac{d\mathbf{v}}{dt} = X_{\mu}(\mathbf{v}), \tag{16}$$

where X_μ is a suitable vector field on a finite-dimensional manifold M, the nature of which can be expressed by very general assumptions. The subscript μ corresponds to an experimental control parameter (for example, Rayleigh number or Reynolds number) ranging over some domain in a control space V. The fluid motion is expected to become irregular and chaotic when μ increases; more precisely, varying μ changes, in general, the structure of the *attracting set* (see below) for the time evolution. For example, it may change from a discrete set of equilibrium points to an attracting limit cycle, and then to an irregular set where the motion is irregular and chaotic. The attracting set thus changes its qualitative nature when μ crosses sets of V called *bifurcation* sets. Therefore, passing through some *scenario* (i.e. bifurcation sequence), one can get turbulent behavior. This kind of study on the possible roads to turbulence has its roots in the general study of deterministic differential equations (*bifurcation theory*), and it constitutes the so-called *geometric theory of chaos*.

In this context one supposes that for some values of the parameter μ the dynamical behavior of the system is well understood. Generally, the system has a stable equilibrium state represented by a fixed point attractor in the phase space M. The aim of the geometric theory of chaos is, then, to give some predictions of the following form: if the attractor undergoes some qualitative changes as μ is varied, then certain other changes are likely to happen as μ is varied further. The precise meaning of 'likely' depends in general on the scenario.

At present we are far from any complete classification of possible scenarios, but at least three of them, leading to nontrivial attractors through qualitative modifications of the equilibrium state, are reasonably well understood (see Eckmann, 1981):

(1) Ruelle–Takens scenario through quasiperiodicity;

(2) Feigenbaum scenario through period doubling;

(3) Pomeau–Manneville scenario through intermittency.

Anyway, turbulence remains a very difficult topic; the fact that we know that in a turbulent flow there is chaos, in the sense discussed above, does not constitute a theory of turbulence, but, at least, it shows that the construction of such a theory necessitates the consideration of chaos, otherwise the theory is doomed to be incorrect. That eliminates some general approaches, for example those which start with stochastic equations or those which consider purely quasiperiodic motions. Indeed, this picture gives a justification of turbulence and some insight into its meaning: there is evidence that the problem of how a fluid changes over from smooth to turbulent flow can be studied through its relation to simple deterministic mathematical systems with similar qualitative behavior. The feature common to these systems is that, as some external parameter is varied, the asymptotic behavior changes from simple to erratic (often through one of the scenarios mentioned above). In the latter situation we observe very often an extreme sensitivity to initial conditions and, consequently, an unpredictable nature of the time evolution.

At present, a relatively large class of systems exhibiting sensitivity to initial conditions is known. Included in this class are, among others, the first-order equations of Lorenz, (9), those of Rössler, the one-dimensional quadratic maps discussed by May, the mapping of the plane into itself investigated by Hénon.

If a time evolution is abstractly described by a flow in a suitable 'phase space', its dynamical properties will be expressed geometrically. In particular, those features which are related to chaoticity arise when the orbits in the phase space converge to an object which is neither a fixed point nor a limit cycle, but rather a strange attractor. The first numerical example of this kind of geometrical objects (see Fig. 5) was given in 1963 by Edward Lorenz. In 1967 Steven Smale discussed strange attractors in the context of abstract dynamical systems, and, in 1970, David Ruelle and Floris Takens have pointed out the strong similarity in the behavior of turbulent flows and strange attractors, suggesting that turbulence results from a strange attractor regime in the Navier–Stokes equation (see Ruelle and Takens, 1970). Moreover, this regime is reached through a scenario (the first mentioned above) involving a sequence of Hopf bifurcations: from a fixed point to a periodic orbit, then to a quasiperiodic one (lying on a two-dimensional torus), and then, after the third bifurcation, it is 'likely' that the system possesses a strange attractor with sensitivity to initial conditions. This scenario is confirmed by some experiments with real fluids, where the power spectra exhibit a broad band when the third independent frequency is about to appear (see Fig. 2).

To conclude our discussion on turbulence, let us come back to the question posed at the beginning about the interpretation of the signal in Fig. 1. In a real experiment the noise present in a signal is usually considered to be the result of the interplay of a large number of degrees of freedom over which one has no control. This type of noise can be reduced by improving the experimental apparatus. But we have seen that another type of noise, which is not removable by any refinement of technique, can be present. This is what we have called the *deterministic noise*. Despite its intractability it provides us with a way to describe noisy signals by simple mathematical models, making possible a dynamical system approach to the problem of turbulence.

Now follows an example of a two-dimensional mapping which has a well known strange attractor.

3

The Hénon mapping

Consider the discrete time evolution equation:

$$x(t+1) = y(t) + 1 - a[x(t)]^2$$
$$y(t+1) = bx(t). \tag{17}$$

Hénon has found this equation by looking for a system which is as simple as possible, yet exhibits the same essential properties as the Lorenz system for certain values of the parameters of that system (not those of Fig. 5) (see Hénon, 1976). The motivation for this research was to provide a more handy model for numerical explorations. The fact that the mapping (17) is discrete in time and two-dimensional, while the Lorenz system is continuous in time and three-dimensional, is because (17) is obtained as a first return map for a surface of section; namely, it is a mapping obtained considering the successive intersections of the three-dimensional flow with a surface Σ of codimension one and transversal to the direction of the flow. Such a mapping is often called Poincaré map. A trajectory is thus replaced by a discrete set of points in Σ, and all the essential properties of the trajectory are carried into corresponding properties of this set of points in such a way that one can forget about the differential system and focus the attention on the mapping T of Σ into itself. Then one can define an explicit equation giving directly $T(x_0, y_0)$ when the initial point with coordinates (x_0, y_0) on Σ is known. Eq. (17) is obtained in such a way and simulates the stretching in one direction and folding over which is a typical behavior with (9). Note that the Hénon dynamics (Fig. 6) is

Fig. 6. The Hénon dynamics: contracting of volume, stretching of length and folding.

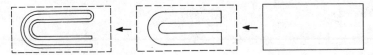

different from Smale's horseshoe dynamics (see Smale, 1967) which does not correspond to an attractor.

Let us notice that the mapping (17) has a constant jacobian:

$$J = \frac{\partial(x(t+1), y(t+1))}{\partial(x(t)), y(t))} = -b. \tag{18}$$

The geometrical interpretation of (18) is that the action of (17) contracts areas by a factor b and reverses the orientation if $0 < b < 1$ (note that an honest Poincaré map would preserve orientation, so that the Hénon map for the usual values $a = 1.4$, $b = 0.3$ cannot really be interpreted as Poincaré map!) We can observe this behavior in Fig. 7a, where the quadrilateral $ABCD$ (which is called the trapping

Fig. 7. The Hénon attractor (see text for details). From Curry (1979) and Hénon (1976).

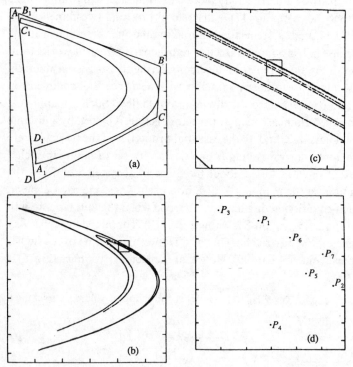

region) is mapped into $A_1B_1C_1D_1$ by the diffeomorphic map of the plane into itself:

$$(x,y) \rightarrow (y + 1 - ax^2, bx). \tag{19}$$

Fig. 7b shows the result of a numerical experiment where an initial point is evolved iterating (17) 10 000 times, with $a = 1.4$ and $b = 0.3$. We can see that the successive points distribute themselves on a complex system of lines which is entirely contained in the trapping region $ABCD$ of Fig. 7a. A magnification of the little square in Fig. 7b yields Fig. 7c, and a further magnification would yield again a similar picture. This system of lines constitutes a strange attractor. The self-similarity visible in Fig. 7b, c is a typical property of a set which is invariant under time evolution, and makes it a *fractal* set (see Mandelbrot, 1982). In the case of the Hénon attractor this is due to its transversal structure (across the lines), which is Cantor-like. If we take $a = 1.3$, $b = 0.3$, the strange attractor of Fig. 7b is replaced by the attracting periodic orbit (with period 7) of Fig. 7d.

4

Capacity and Hausdorff dimension

Roughly speaking the dimension of a set is the amount of information needed to specify points in it accurately. In general, describing the size of a subset of R^m implies naturally the use of Lebesgue measure. However, in the study of nonconservative systems we are very often interested in invariant sets which are of Lebesgue measure zero. In the case of the Hénon mapping we have a situation in which an open neighborhood $U \subset R^2$ is mapped into itself, with $|J| < 1$ (see (18)). Since the volume is shrunk with each application of the map, the asymptotic set shown in Fig. 7b must have Lebesgue measure zero.

Moreover, we have seen that on a strange attractor there are directions which are expanded by the map.

We shall now define some indicators (dimensions) which make it possible to compare two sets of Lebesgue measure zero to see which one is 'larger'.

Let A be a nonempty set with a metric, and $N(r,A)$ the minimum number of open balls of radius r needed to cover A. Then, the *capacity* (in the Kolmogorov terminology) is given by:

$$\dim_K(A) = \lim_{r \to 0} \sup \frac{\log N(r,A)}{\log (1/r)}. \tag{20}$$

Fig. 8 gives an example of a very simple fractal object: the Cantor set.

In this simple case we have:

$$\lim_{r \to 0} \frac{\log N(r)}{\log (1/r)} = \lim_{n \to \infty} \frac{\log 2^n}{\log 3^n} = \frac{\log 2}{\log 3} < 1. \tag{21}$$

Another important concept, introduced in 1919, is the *Hausdorff dimension*.

Consider the definition (14) and (15) of Hausdorff measure in dimension α. The Hausdorff dimension of A is defined by considering the behavior of $m^\alpha(A)$ not as a function of A, but as a function of α. It is easily seen that there exist a point α_0 such that $m^\alpha(A) = +\infty$ for $\alpha < \alpha_0$ and $m^\alpha(A) = 0$ for $\alpha < \alpha_0$. The uniquely defined number α_0 is the Hausdorff dimension of A, and we denote it by $\dim_H(A)$. We have

$$\dim_H(A) = \sup\{\alpha : m^\alpha(A) = +\infty\} = \inf\{\alpha : m^\alpha(A) = 0\}. \quad (22)$$

Two basic properties of Hausdorff dimension are the following:

$$\dim_H(A) \le \dim_H(A') \text{ if } A \subset A'$$

and

$$\dim_H(\cup_n A_n) = \sup_n \dim_H A_n.$$

Finally, for every compact set A, we have

$$\dim_H(A) \le \dim_K(A).$$

Now, let A be a subset of the unit interval $[0,1]$. For such a set we have that $m^1(A) \le 1$ because m^1 is just the ordinary outer measure (see (15)). Thus $\dim_H(A)$ must lie between zero and one. In the case that A is a Borel set of positive Lebesgue measure (a segment for instance), we get $m^1(A) > 0$ and hence $\dim_H(A) = 1$. Conversely, every one-point set, and hence every countable set, has Hausdorff dimension zero. This shows that ordinary geometrical objects get the dimension they should. In between we can find, for example, the Cantor set, for which we have obtained the dimension $\log 2/\log 3$ (for this simple fractal set, Hausdorff dimension and capacity coincide).

Note that B. Mandelbrot has introduced the convenient phrase 'fractal dimension' to denote Hausdorff dimension, but that other authors mean by that the Kolmogorov capacity.

Here we have introduced some *geometrical concepts*. In the next section we will discuss some *statistical* concepts concerning the properties of the *probability measures* ρ carried by an attractor. An

Fig. 8. The construction of a Cantor set up to the second step $n = 2$.

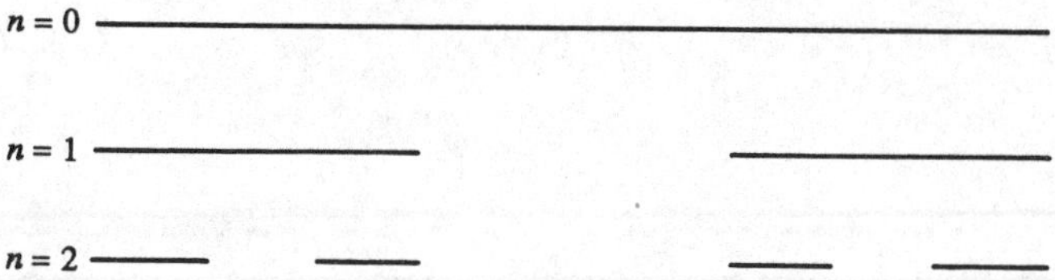

important quantity, which is often computed in practical cases, is the *information dimension* $\dim_H \rho$, which is defined as the minimum of the Hausdorff dimension of the sets A for which $\rho(A) = 1$. From here on, we shall call a fractal set a set for which the Hausdorff dimension (or the information dimension) is not an integer. In the case of the Hénon attractor with $a = 1.4$ and $b = 0.3$, the value 1.21 is found for the information dimension (Grassberger, 1986).

Coming back to Fig. 7, if one repeats the operation done in Fig. 7b taking $a = 1.3$, $b = 0.3$, the strange attractor disappears and is replaced by a periodic orbit of period 7, Fig. 7d. Now, if we consider two initial points which are close to each other, at distance, say, ε_0, then from numerical experiments we find that in the situation of Fig. 7b the system has sensitive dependence on initial conditions:

$$\varepsilon_t \sim \exp(\lambda t)\varepsilon_0, \tag{23}$$

where $\lambda = 0.42$ is the largest *characteristic exponent*. The same thing repeated for $a = 1.3$ would reveal no chaoticity at all, λ being negative.

5

Attracting sets and attractors

As we have seen in (18), the Hénon mapping contracts areas by a factor b, and therefore constitutes a *dissipative system*. Conversely, for a *conservative system*, i.e. a Hamiltonian time evolution, Liouville's theorem says that the volume in phase space is conserved. Generally, we define a *differentiable dynamical system* by an evolution equation of the form (continuous case)

$$\frac{d\mathbf{x}}{dt} = F(\mathbf{x}) \tag{24}$$

or by a map (discrete case)

$$\mathbf{x}(t+1) = f(\mathbf{x}(t)), \tag{25}$$

where F and f are differentiable functions and the variables $\mathbf{x}$ vary over a phase space M, which can be R^m or a compact manifold, or infinite dimensional space. The examples encountered so far cover all these possibilities: the Lorenz system is of the form (24) and M is R^3; the Hénon mapping is of the form (25) and M is R^2; the Navier–Stokes equation is of the form (24) and M is the (infinite-dimensional) Hilbert space H of square-integrable vector fields which are orthogonal to gradients; finally the geodesic flow considered by Hadamard is of the form (24) and M is the unit tangent bundle of a surface with constant negative curvature. Then, we introduce the nonlinear time evolution operators f^t, with the property:

$$f^0 x = x \text{ and } f^{t_1 + t_2} x = f^{t_1} \circ f^{t_2} x \tag{26}$$

so that f^t is a group or (if $t \geq 0$) a semi-group on M. In the following we shall be mainly interested in dissipative systems, where, in general, one can assume that there is a set U in phase space M which is contracted by time evolution asymptotically to a compact set $A = \cap_{t \geq 0} f^t U$.

In the case of the Hénon system (17), U can be identified with the quadrilateral $ABCD$ in Fig. 7a and A is presumably represented by the system of lines showed in Fig. 7b. For the Lorenz system one can consider a ball $U = \{(x,y,z) : x^2 + y^2 + z^2 \leq R^2\}$ with large R, then U is mapped into itself by time evolution, and A is the closure of the set of points visible in Fig. 5.

Generally, the set A will be called an *attracting set* with fundamental neighborhood U, if it satisfies the following properties (see Ruelle, 1981).

(1) *Attractivity:* for every open set $V \supset A$ we have $f^t U \subset V$ for all sufficiently large t.

(2) *Invariance:* $f^t A = A$, for all t.

These two conditions define an attracting set. Moreover, its *basin of attraction* is defined to be the set of initial points x such that $f^t x$ approaches A as $t \to \infty$, i.e. $\bigcup_{t < 0} f^t U$.

We shall not at this point give a mathematical definition of an *attractor*, but only an operational definition, saying that it is the set on which experimental points $f^t x$ accumulate for large t. An attractor A should again be invariant: $fA = A$, but need not have an *open* basin of attraction. This means that an attractor need not be stable under noise of finite amplitude. On the other hand, we would like an attractor to be *irreducible* in some sense:

(3) *Irreducibility:* if an attracting set consists of a number of disjoint invariant pieces, it may happen that some of them are not attracting. Then we consider as an attractor each irreducible piece which is indeed attracting. An irreducibility condition could be the following: one can choose a point $x' \in A$ such that for each $x \in A$ there is a positive t such that $f^t x'$ is arbitrary close to x (topological transitivity).

Note that it is not known if the attracting set for the Hénon system is topologically transitive: there may thus be a difference between attracting set and attractor in this case.

The property which makes an attractor *strange* is the one already discussed:

(4) *Sensitive dependence on initial conditions:* the notion of strangeness is thus related to dynamical properties of an attractor, and not just to its geometry. More precisely, let us say that a strange attractor is defined by the fact that its *asymptotic measure* (see below) has a positive characteristic exponent.

Both the Hénon mapping and the Lorenz equations have attractors which are strange in this sense, but there are examples of attractors with a complicated geometrical structure which are not strange. One of them is the Feigenbaum attractor which arises from the logistic map of the interval [0,1] into itself (May, 1976):

$$x(t+1) = \mu x(t)(1 - x(t)). \tag{27}$$

Varying the parameter μ over [0,4], this simple map will go through a whole spectrum of possible dynamical behaviors. In particular, when $\mu < 3$ it has an attracting fixed point which becomes unstable at $\mu = 3$. For $\mu > 3$ the map (27) has an attracting periodic orbit of period 2^n, with n tending to infinity as μ tends to 3.57. . . . When the latter value is reached there is the attractor shown in Fig. 9.

It is found that this attractor has a fractal structure but is not chaotic. Moreover, let us notice that this is an attractor but not an attracting set because it is a limit of repelling periodic points. This shows that an attractor is in some sense a 'weaker' mathematical object than an attracting set.

As we have already pointed out, the presence of chaos implies a strong sensitivity to small fluctuations. Some numerical studies carried out by Curry (1979) have shown that if one computes the orbit of (17) (with $a = 1.4$) with two different computers, since they have different round off errors the orbits which are computed are not

Fig. 9. The asymptotic measure on the Feigenbaum attractor plotted as a histogram of 100 000 points in 1000 bins.

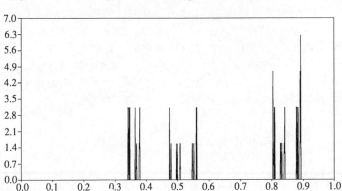

exactly the same. Suppose they work with 14 digits of precision, then Curry found that after only 60 iterations the difference between the outputs of the two computers becomes of order one, that is the size of Fig. 7a. Hence, we can ask why we should compute 10 000 points if after 60 the results are already meaningless. *The point is that we have to reformulate what is considered as meaningful.* In fact, it is clear that if one does not make more precise computations, the exact position of the 60th point of the orbit is *completely indeterminate*: it could be anywhere on the attractor; however, as far as the *statistical* properties of the orbit are concerned, this must not trouble us because they do not depend on the details of the calculations and therefore the existence of the attractor is not in question. This is also important if we are interested in the study of experimental situations where one expects that the equations governing time evolutions are always affected by some noise. *The properties which are investigated should be stable under small fluctuations.*

In particular, small fluctuations play an essential role in defining attractors. In fact, the latter should have the further property:

(5) *Stability under small random perturbations*: we require that the motion of a dynamical system submitted to small random perturbations (like the round off errors due to the floating point truncation in computer operations, or the external noise in experimental situations) be as asymptotically concentrated on attractors, and that the asymptotic measure ρ be stable under such perturbations (see below).

So attractors, as experimental objects, give us a global description of the asymptotic behavior of dynamical systems. Moreover, these objects come equipped with an asymptotic measure, namely a probability measure ρ on A, which describes how frequently various parts of A are visited by the orbit. An example is the histogram shown in Fig. 9. Such a measure represents the so-called *physical measure*, namely the measure ρ describing time averages:

$$\rho(\phi) \equiv \int \rho(\mathrm{d}x)\phi(x) = \lim_{T \to \infty} \frac{1}{T} \int_0^T \mathrm{d}t\,\phi[x(t)]. \tag{28}$$

We refer to part II for the discussion of invariant probability measures corresponding to time averages. Let us say that, once again, the random noise which is usually present in experimental studies can provide the selection of the physical measure in a very natural way. In fact, due to the presence of noise, a physical time evolution can be considered as a stochastic process, which has normally only one stationary measure ρ_ε. Then, we could define the asymptotic measure ρ (called sometimes *Kolmogorov measure*) as the 'zero-noise' limit ($\varepsilon \to 0$) of ρ_ε. We shall see how this claim is substantiated in the case of some particular dynamical systems called *Axiom A systems*.

6

Extracting geometric information from a time series

To conclude this first part we want to discuss briefly the problem of reconstructing phase space pictures from the observation of a single coordinate $x(t)$, monitored from experimental situations.

The general philosophy of this approach is to extract 'physical sense' from an experimental signal, bypassing the detailed knowledge of the underlying dynamics. In particular, one investigates *some* properties which characterise the time evolution in terms of geometrical quantities.

The general technique of this approach is to generate several different scalar signals $x_k(t)$ from the original $x(t)$ in such a way as to reconstruct an N-dimensional space where, under some conditions, we can obtain a good representation of the attractor.

The easiest way to do that is to use *time delays* (originally, time derivatives were used: $x_1(t) = x(t)$, $x_2(t) = \dot{x}(t)$, $x_3(t) = \ddot{x}(t)$ ... then Ruelle proposed using time delays, this proposal was not published but is recorded in Packard *et al.* (1980, footnote 8). We write:

$$x_k(t) = x(t + (k-1)\tau) \text{ with } k = 1 \ldots N. \tag{29}$$

In this manner, an N-dimensional signal is generated, which can be represented by the vector:

$$\mathbf{x}(t) = \begin{pmatrix} x(t) \\ x(t+\tau) \\ \vdots \\ x(t+(N-1)\tau) \end{pmatrix}. \tag{30}$$

Note that, on varying the set of variables which can be constructed from $x(t)$, we get in principle the same geometric information.

In Fig. 10 we show the Rössler attractor reconstructed with two different sets of variables. The dynamical system generating this attractor is the set of three ordinary differential equations:

$$\frac{dx}{dt} = -(y+z)$$

$$\frac{dy}{dt} = x + 0.2y \tag{31}$$

$$\frac{dz}{dt} = 0.4 + xz - 5.7z.$$

This dynamical system arises from modelling a chemical reaction (see below). A comparison of the two pictures indicates that the qualitative characteristics of the attractor remain intact when the coordinates are changed.

Next, the reconstruction of a 'true' attractor A which lives in a usually infinite-dimensional space M, is nothing but an N-dimensional projection πA, the shape of which depends on the number N of variables and on the delay τ.

Hence, the dimension of the attractor A should be small enough to be tractable. There are some theorems (Mañé, 1981; Takens, 1981) which state that, in order to obtain a good projection πA (without trajectories crossing each other), N must be about twice the Hausdorff dimension of A.

Maybe the best example up to now of this procedure comes from

Fig. 10. (a) The projection of the 'true' Rössler attractor, plotted with two phase-variables (x,y). (b) Its reconstruction from the time series, in the plane $(X,dx/dt)$. From Packard *et al.* (1980).

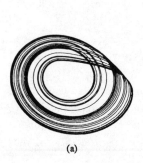
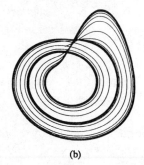

(a)   (b)

chemical kinetics. Let us point out that chaos may also be observed in this field thanks to the nonlinearity of the kinetic mass action law.

A typical experimental situation is to have an open reactor continuously fed with some initial reactants, and provided with a suitable overflow exit. Inside the reactor the chemicals are rapidly mixed by a stirring device. This is what in chemical engineering is called CSTR (continuously stirred tank reactor).

If we denote the jth chemical species as X_j (as well as its concentration) and as X_j^0 the concentration of X_j in the inlet flow, then the time evolution of the chemical reaction is given by:

$$\frac{\mathrm{d}X_j}{\mathrm{d}t} = F_j(X_1, \ldots, X_n) + \mu(X_j^0 - X_j) \text{ with } j = 1 \ldots n. \qquad (32)$$

Here the control parameter μ represents the inverse of the mean residence time in the reactor, and it is easily controllable; the nonlinear terms F_j describe the kinetic of the reaction.

A chemical reaction taking place in a CSTR is thus a typical example of a system governed by an evolution equation of the form (1), exhibiting a bifurcation parameter μ which can be kept constant in each run of the experiment. It has been known for some time that for certain reactions, like the famed Belousov–Zhabotinsky reaction, a chemical system may undergo self-sustained oscillations which can be periodic or chaotic.

Fig. 11a displays a time series for the concentration of a reactant in a parameter regime where there is no periodicity over long records (many hours, the period of a single oscillation being some seconds). In Fig. 11b we see the attractor reconstructed by the method of time delays. If we cut the attractor along a Poincaré section (the transversal dotted line in Fig. 11b), we obtain the set of points plotted in Fig. 11c. Then, given a point x in the Poincaré section, we can construct the *first return map*, namely the map which will bring x to Px, again in the Poincaré section. The first return map, plotted in Fig. 11d, gives information about the nature of the observed chaos. The fact that actually the points in Fig. 11d fall along a well defined curve indicates the deterministic nature of this chaotic time evolution.

Whenever a good model of P can be constructed, one has

essentially succeeded in extracting some physical sense from the experimental time series.

However, this procedure of 'interpretation' of a chaotic signal is possible only for those systems for which precise geometrical information about the shape of the attractor is accessible. In other words, since it is only feasible to model geometrically a time evolution with relatively few variables, this procedure is restricted to those cases in which one has low-dimensional attractors.

In order to extract useful information from a time evolution which lives on a relatively high-dimensional attractor, we have to put

Fig. 11. (a) The time series $x(t)$ of bromide ion concentration in the Belousov–Zhabotinsky reaction. (b) Plot of the reconstructed attractor in the plane $(x(t),x(t+\tau))$. (c) Poincaré section of the attractor along the cut. (d) First return map of the Poincaré section. From Roux and Swinney (1981).

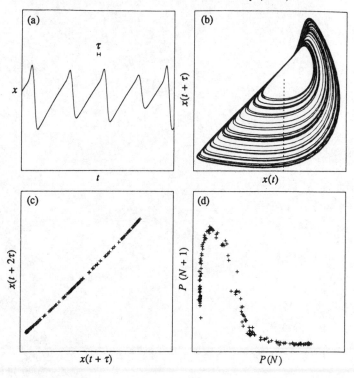

ourselves into the context of ergodic theory. In this context we study properties of the invariant probability measure ρ generated by the *density* of points $f^t x$ in the phase space M as the time t goes to infinity.

Moreover, shifting the attention from attractors to invariant measures we are able to study systems which are excited beyond the simplest bifurcations, and thus we move toward a better understanding of what happens into the gap dividing simple chaotic systems and fully developed turbulence.

Part II   The ergodic theory of chaos

7

Invariant probability measures

As we have already mentioned, a differentiable dynamical system on a compact manifold M (or other kinds of spaces, it does not matter very much for our purposes) is a family of differentiable evolution operators (or maps) $f^t : M \to M$, such that $f^0 =$ identity and $f^{t_1 + t_2} = f^{t_1} \circ f^{t_2}$.

The time t may vary over the reals (continuous case) or the integers (discrete case), possibly with the restriction $t \geq 0$, in such a way that f^t operates as a semi-group on M.

The essential fact is the existence of a *probability measure ρ* on M, which is invariant under the time evolution f^t.

There are several ways of presenting measure theory. In our context, one notion which is particularly appropriate is that of *Radon measure*.

Given a continuous function $\phi: M \to R$ consider the integral:

$$\rho(\phi) \equiv \int \phi(x)\rho(\mathrm{d}x). \tag{33}$$

The notion of Radon measure is that one can *define* the measure ρ by the integral (33). More precisely, let ϕ be a function ranging over the class $C^0(M)$ of continuous functions on a compact metric space M. We define a *continuous linear functional* as an assignment $E: C^0 \to R$ satisfying the addition rule (linearity):

$$E(\alpha_1 \phi_1 + \alpha_2 \phi_2) = \alpha_1 E(\phi_1) + \alpha_2 E(\phi_2), \tag{34}$$

and, if $\{\phi_n\}$ is a sequence of functions in $C^0(M)$ converging uniformly to zero, the (continuity):

$$E(\phi_n) \to 0. \tag{35}$$

We say that any continuous linear functional on $C^0(M)$ is a measure on M (Radon measure).

In fact, we shall be interested only in positive measures, i.e. linear functionals such that $E(\phi) \geq 0$. For such functionals, continuity is automatically satisfied. A *probability measure* is a positive measure such that $E(1) = 1$, where $1 \in C^0(M)$ is the constant function 1. We write now

$$E(\phi) = \rho(\phi) \equiv \int \phi(x) \rho(\mathrm{d}x). \tag{36}$$

Thus, if ρ is a probability measure, we have $\rho \geq 0$, $\rho(1) = 1$.

As far as a dynamical system f^t on M is concerned, a probability measure ρ on M defines an *expectation value*:

$$\langle \phi \rangle = \rho(\phi), \tag{37}$$

which, operationally, is provided by the *time average*:

$$\rho(\phi) \equiv \int \phi(x) \rho(\mathrm{d}x) = \lim_{T \to \infty} \frac{1}{T} \int_0^T \mathrm{d}t \, \phi(f^t x). \tag{38}$$

Now, given a probability measure ρ, one can define the notions of *invariance* and *ergodicity*.

Invariance: $\rho(\phi) = \rho(\phi \circ f^t)$, where $(\phi \circ f^t) x = \phi(f^t x)$. This means that the measure ρ is invariant under time evolution.

In the context of abstract ergodic theory one studies the transformations which preserve the structure of a probability space $(M, \mathcal{R}, \rho)$. Namely those transformations $f^t : M \to M$ which are *measurable* (i.e. $A \in \mathcal{R}$ implies $f^{-t} A = \{x : f^t x \in A\} \in \mathcal{R}$) and *measure-preserving*:

$$\rho(A) = \rho(f^{-t} A). \tag{39}$$

The reason for which in (39) we use f^{-t} instead of f^t is that the map f^t may not be invertible.

Example 1

Let M be the half-open unit interval $[0,1)$ and $\mathcal{R}$ consisting of the Borel subset of M (i.e. $\mathcal{R}$ is the σ-algebra generated by intervals). Consider the map

$$f(x) = 2x \pmod{1}, \tag{40}$$

that is

$$f(x) = \begin{cases} 2x, & \text{if } x \in [0, \frac{1}{2}) \\ 2x - 1, & \text{if } x \in [\frac{1}{2}, 1) \end{cases}. \tag{41}$$

This map can be considered as a map $f: S^1 \to S^1$ of the unit circle into itself by a parametrisation of points in S^1 by numbers in the interval $[0,1)$. On the other hand, a number $x \in [0,1)$ can always be represented by a binary expansion $0.\omega_1\omega_2\omega_3 \ldots$ (which terminates if x is a dyadic rational), where, for each n, $\omega_n = 0$ or 1. It is clear that the map (40) replaces $0.\omega_1\omega_2\omega_3 \ldots$ by $0.\omega_2\omega_3 \ldots$: the most significant digit has been dropped reflecting the fact that this map is not invertible, being two-to-one.

This kind of operation on a string of symbols is called a *shift*. In general, we call a shift a transformation $\sigma: \Omega \to \Omega$, where the general element of Ω is a doubly infinite sequence of the form $\omega = (\ldots \omega_{-1}, \omega_0, \omega_1, \ldots)$ of elements of a finite set V and σ operates on an element ω in such a way that $(\sigma\omega)_n = \omega_{n+1}$.

Let us assign a probability p_i to each element i of V (i.e. $p_i \geq 0$, $\Sigma_{i \in V} p_i = 1$). We obtain a particular probability measure ρ on Ω by requiring that probability$(\omega_n = i) = p_i$ independently for each n. This measure is known as Bernoulli measure on Ω (one also speaks of Bernoulli shift for the pair (ρ, σ)). Clearly, ρ is invariant under the shift. For the case of the shift which corresponds to the map (40), Ω consists of half-infinite rather than doubly infinite sequences (but this makes little difference), and $V = \{0, 1\}$. Consider the Bernoulli measure such that probability$(\omega_n = 0) = p$, probability$(\omega_n = 1) = 1 - p$; there are uncountably many such measures, corresponding to different choices of p in $(0, 1)$.

Looking at Fig. 12 we can see that the Lebesgue measure m, which

Fig. 12. The dyadic transformation. The Lebesgue measure m is invariant for this map: $m(f^{-1}(I)) = m(I_1) + m(I_2) = m(I)$.

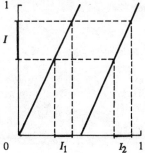

corresponds to the choice $p = \frac{1}{2}$, is preserved under the iteration of (40). This reflects the fact that the map (40) together with the Lebesgue measure and the Bernoulli shift with $V = \{0,1\}$ and $p = \frac{1}{2}$, denoted by $\{\frac{1}{2}, \frac{1}{2}\}$, are measure-theoretically conjugated (see, for instance, the excellent book by Billingsley, 1965, for a discussion of such matters).

Ergodicity: An invariant probability measure ρ is *ergodic* or *indecomposable* if it does not have a nontrivial convex decomposition:

$$\rho = \alpha \rho_1 + (1 - \alpha) \rho_2 \text{ with } \alpha \neq 0, 1, \tag{42}$$

where ρ_1 and ρ_2 are again invariant probability measures and $\rho_1 \neq \rho_2$. Note that if ρ is ergodic, and A is an invariant set, then $\rho(A) = 0$ or 1.

There is a theorem which states that if a compact set A is invariant under the time evolution f^t, where the maps $f^t: A \to A$ are continuous, then there is a probability measure ρ invariant under f^t and with support contained in A; moreover, one may choose ρ to be ergodic.

Generally, if we have a compact metric space M and a time evolution f^t on it, then there are always invariant probability measures which can be chosen to be ergodic, and any invariant probability measure ρ can be represented in a unique way as a convex sum (or integral) of ergodic measures (ergodic decomposition).

Let us make the point more precise. Given an initial point $x_0 \in M$, we say that the *orbit* $\{f^i x_0\}_{i \geq 0}$ is *statistically regular* if the time average

$$\bar{\phi}(x_0) = \lim_{n \to \infty} \frac{1}{n} \sum_{i=0}^{n-1} \phi(f^i x_0) \tag{43}$$

exists for all continuous functions ϕ. Then there is a unique probability measure ρ_0 such that

$$\int \phi(x) \rho_0(\mathrm{d}x) = \bar{\phi}(x_0). \tag{44}$$

Now, it may happen that $\bar{\phi}$ is not the same depending on the region considered, namely we can find an initial point $\hat{x}_0 (\neq x_0)$ with an orbit which is also statistically regular but nevertheless $\bar{\phi}(\hat{x}_0) \neq \bar{\phi}(x_0)$.

Let us now state the ergodic theorem.

Ergodic theorem

Let ϕ be an integrable function (for our purposes ϕ may be chosen as continuous) and ρ an invariant probability measure, then for ρ-almost all x_0 the limit (43) exists. If ρ is ergodic then $\bar{\phi}$ is almost everywhere constant, and this constant is given by (44).

If ρ is not ergodic, there corresponds via (43) an ergodic measure ρ_0 to ρ-almost every x_0: these ρ_0 are those which occur in the ergodic decomposition of ρ.

The notion of ergodicity arose in the context of statistical mechanics. In particular the works of Boltzmann and Gibbs raised the so-called *ergodic problem*: to find conditions under which the continuous time analog of the limit (43) exists and is almost everywhere constant with respect to the Liouville measure ρ. In 1931 Birkoff proved the ergodic theorem showing that a necessary and sufficient condition for the ergodic problem is that there exists no set A invariant under time evolution such that:

$$0 < \rho(\chi_A) < 1, \tag{45}$$

where χ_A is the characteristic function of the set A. Nevertheless Birkoff's result did not close the problem, since for the time evolutions which occur in statistical mechanics, ergodicity remains, in general, a property which is difficult to prove.

Note that it does not make sense to say that f^t is ergodic if you do not say with respect to what measure ρ. In the case of a conservative system there is a smooth invariant measure m given by the Liouville theorem, and we say that f^t is ergodic if m is an ergodic measure with respect to f^t. If f^t is a dissipative system, one has to specify which ergodic measure ρ one considers, such a measure is expected to be concentrated on an invariant set of Lebesgue measure zero.

Let us point out that in the statistical analysis of time series we deal directly with time averages like (43) and not with an invariant probability measure ρ given *a priori*. Therefore the investigation will be concentrated, in general, on the properties of an ergodic measure ρ *defined* by the time series. Statistical analysis of chaos has not very much to do with the ergodic problem. Actually it 'bypasses' this problem, focussing its attention on invariant measures carried by attractors, which are *assumed* to be ergodic.

To conclude this section we give an example of a time evolution which has a naturally defined ergodic invariant measure.

Example 2

In chapter 5 we discussed some properties of the mapping of the interval [0,1] into itself $x \to \mu x(1-x)$, and we have seen that when $\mu = 3.57 \ldots$ the time evolution lives on an attractor whose invariant density is shown in Fig. 9. In that case the invariant set A on which the density is concentrated is of Lebesgue measure zero and the ergodic invariant measure ρ is clearly singular with respect to the Lebesgue measure.

Now consider the case $\mu = 4$:

$$f(x) = 4x(1-x). \tag{46}$$

This map was studied first by S.M. Ulam and J. von Neumann (1947), and it is the type of a deterministic map with a completely random behavior, as we have already seen in Fig. 1. If we plot the density of points as we have done for the previous case (Fig. 9), we obtain the histogram shown in Fig. 13.

Moreover, for this particular case we can actually calculate the ergodic invariant measure ρ simply by a change of coordinates (see Collet and Eckmann, 1980b).

Fig. 13. The invariant density for the map (46) plotted as a histogram of 100 000 points in 1000 bins.

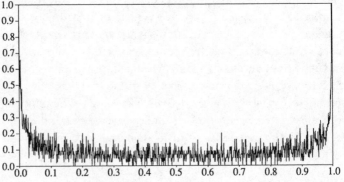

Let ϕ be the homeomorphism of [0,1] into itself given by:

$$\phi(x) = \frac{2}{\pi}\arcsin x^{\frac{1}{2}}. \tag{47}$$

Then, we have at once that $\tilde{f}(x) = \phi \circ f \circ \phi^{-1}$ is the 'tent' map, given by

$$\tilde{f}(x) = \begin{cases} 2x & \text{if } x \in [0, \frac{1}{2}] \\ 2 - 2x & \text{if } x \in [\frac{1}{2}, 1]. \end{cases} \tag{48}$$

The map (48) preserves the Lebesgue measure m exactly in the same way as the map discussed in Example 1. Hence, f preserves the measure $\rho = \phi^{-1}m$, namely:

$$\rho(dx) = \frac{d\phi}{dx}dx = \frac{1}{\pi}\frac{dx}{[x(1-x)]^{\frac{1}{2}}}. \tag{49}$$

The measure ρ is an example of a measure which is absolutely continuous with respect to the Lebesgue measure. It is easy to see that any absolutely continuous invariant measure ρ for a map $f: R^n \to R^n$, can be written as:

$$\rho(dx) = h(x)m(dx) \equiv h(x)dx, \tag{50}$$

where the density $h(x)$ is in $L^1(m)$ and satisfies the equation:

$$h(y) = \sum_{f(x)=y} \frac{h(x)}{|\det[Df(x)]|}. \tag{51}$$

(see Fig. 14). From (51) we have that Lebesgue measure is invariant

Fig. 14. The graph of the density $h(x) = 1/\pi[x(1-x)]^{\frac{1}{2}}$ normalised for a comparison with the histogram in Fig. 13.

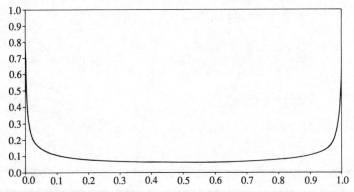

for a one-to-one map f if and only if $|\det(Df)| \equiv 1$ (volume preservation).

In our (one-dimensional) case we have $Df(x) = df/dx$, i.e. the slope of the map. Therefore (51) says that the density of points which fall on a small interval around y is equal to the sum of the densities at the inverse points of the map f weighted by the inverse absolute value of the slope at those points (see Fig. 15).

Then the absolute value of the slope is related to the *spreading* of trajectories. Considering the tent map (48), it is easily seen that an imprecision δx, no matter how small, of our knowledge of a point x will be amplified by a factor two with each iteration, since the derivative of $\tilde{f}$ equals ± 2. The same happens for the map (46) if we consider the *average* behavior.

If we denote the amplification *rate* of f: $[0,1] \rightarrow [0,1]$ at a particular point x by $\log |df/dx|$, then the average rate of growth of errors is given by:

$$\lambda = \int_0^1 \log \left| \frac{df}{dx} \right| \rho(dx). \tag{52}$$

The quantity (52) is called the *characteristic exponent* for the measure ρ. If we compute λ for the map (46) we will find that its invariant measure (49) is not only ergodic but has also a *positive* characteristic exponent, i.e. map (46) is chaotic in the sense we have discussed in part I:

Fig. 15. From Shaw (1981).

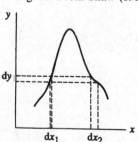

$$\lambda = \frac{1}{\pi} \int\limits_0^1 \frac{\log|4(1-2x)|}{[x(1-x)]^{-\frac{1}{2}}} \mathrm{d}x = \log 2. \tag{53}$$

In the framework of information theory the result (53) means that map (46) is an 'information source' producing exactly one bit per iteration (see Shaw, 1981).

8

Physical measures

In general, it is exceptional that an attractor carries only one ergodic invariant measure ρ. In typical cases there are uncountably many distinct ergodic measures. Nevertheless, as we have already mentioned in part I, in physical experiment and in computer simulations it seems that one invariant probability measure ρ is produced more or less automatically by the time that the system spends in various part of the space M. Thus, there is a selection process of the so-called *physical measure* ρ.

The advantage of the ergodic approach lies in the fact that there are important theorems which apply to *all* ergodic measures, and we do not have to worry immediately about which ergodic measure is physical. (Furthermore, as noted earlier, there are always *some* ergodic measures on a compact invariant set.)

In some cases, we can attempt to construct some selection processes to obtain measures which describe physical time averages. We have seen in chapter 6 that a candidate for this purpose is the *Kolmogorov measure*, obtained by the 'zero-noise' limit $\varepsilon \to 0$ of a stationary probability measure ρ_ε. Another possibility is represented by a so-called SRB measure, which is the measure given by the time average (43) for all x_0 in a set $A \subset M$ with Lebesgue measure $m(A) > 0$. Moreover we shall see that for Axiom-A systems these two measures coincide.

However, many important results of ergodic theory hold for an arbitrary invariant measure ρ. This is the case, for example, of the existence of characteristic exponents.

9

Characteristic exponents

In Example 2 we have seen that the growth rate of Df^n (the derivative of f composed with itself n times), is related to the exponential rate at which nearby orbits are separated. In fact, if we consider the discrete time evolution equation

$$x(n+1)=f(x(n)), \quad x(i)\in R, \tag{54}$$

the separation of two initial points $x(0)$ and $x'(0)$ after time n is given by

$$\begin{aligned} x(n)-x'(n) &= f^n(x(0))-f^n(x'(0)) \\ &\approx [Df^n(x(0))](x(0)-x'(0)), \end{aligned} \tag{55}$$

where, in this case, $Df=df/dx$. Applying the chain rule of differentiation we have

$$Df^n(x(0))=Df(x(n-1))Df(x(n-2)) \ldots Df(x(0)). \tag{56}$$

Therefore we can define the average rate of growth as the number

$$\lambda=\lim_{n\to\infty}\frac{1}{n}\log |D_{x(0)}f^n\delta x(0)|. \tag{57}$$

We shall see that the limit (57) exists for ρ-almost all $x(0)$ in a great generality. This is guaranteed by a generalisation of the ergodic theorem to products of matrices, proved by Oseledec (1968).

9.1   The multiplicative ergodic theorem

Theorem
Let ρ be a probability measure in a space M, and $f: M\to M$ a measure preserving map such that ρ is ergodic. Let also $T: M\to m \times m$ matrices be a measurable map such that:

$$\int\rho(dx)\log^+ \|T(x)\|<\infty, \tag{58}$$

where $\log^+ u=\max(0,\log u)$.

Define the matrix $T_x^n = T(f^{n-1}x) \ldots T(fx)T(x)$. Then, for ρ-almost all x, the following limit exists:

$$\lim_{n \to \infty} (T_x^{n*}T_x^n)^{1/2n} = \Lambda_x, \tag{59}$$

where T_x^{n*} is the adjoint of T_x^n.

The logarithms of the eigenvalues of Λ_x are called *characteristic exponents*, or *Liapunov exponents*. We denote them by $\lambda_1 \geq \lambda_2 \geq \ldots$ or by $\lambda^{(1)} > \lambda^{(2)} > \ldots$ when they are no longer repeated by their multiplicity $m^{(i)}$. The characteristic exponents are almost everywhere constant if ρ is ergodic.

It is clear that in the case $m = 1$, where 1×1 matrices are ordinary numbers, the multiplicative ergodic theorem reduces to the ordinary ergodic theorem. The difficulty, but also the novelty, of the multiplicative ergodic theorem is that when $m > 1$ it deals with *noncommuting* matrices.

Let us say that a point $x \in M$ is a *regular point* of f if R^m can be written as follows:

$$R^m = E_x^{(1)} \supset \ldots, \tag{60}$$

so that

$$\lim_{n \to \infty} \frac{1}{n} \log \| T_x^n \mathbf{u} \| = \lambda^{(i)} \tag{61}$$

whenever $0 \neq \mathbf{u} \in E_x^{(i)} \backslash E_x^{(i+1)}$. Another version of the multiplicative ergodic theorem, which is not immediately equivalent to the previous one, is then the following:

Theorem

With the assumptions of the previous theorem, the set of regular points is a Borel set of full ρ-measure, namely the limit (61) does exist ρ-almost everywhere.

In particular, for all $\mathbf{u} \in R^m$ that are not in the subspace $E_x^{(2)}$ (almost all $\mathbf{u}$), the limit (61) gives the largest characteristic exponent $\lambda^{(1)}$.

The relation between the two forms of the theorem is that the second holds if $E_x^{(i)}$ is defined to be the subspace of R^m spanned by the eigenspaces of Λ_x corresponding to eigenvalues $\leq \exp(\lambda^{(i)})$. In particular, the dimension of $E_x^{(i)}$ is the sum of the multiplicities $m^{(j)}$ of the eigenvalues $\lambda^{(j)} \leq \lambda^{(i)}$.

Notice that the subspace $E_x^{(i)}$ and the characteristic exponents $\lambda^{(i)}$ do not change if we replace the Euclidean norm ‖ ‖ by some other norm on R^m.

Another important remark concerns the inversion of time. If f^t is a diffeomorphism (i.e. f has an inverse f^{-1} which is a differentiable map), we may consider the time-reversed dynamical system f^{-t}. The invariant measure ρ of the original system is also invariant for the time-reversed system, and changing the direction of time (i.e. $t \to -t$) the characteristic exponents are the same but with opposite sign: $\lambda_i \to -\lambda_i$.

In some physical applications, like those concerning hydro-dynamical systems, we shall need a version of these theorems where R^m is replaced by an infinite-dimensional space, like a Banach space or a Hilbert space E and the $T(x)$ are bounded operators (i.e. there exists a positive real number α such that $|T\mathbf{u}| \le \alpha |\mathbf{u}|$ for each $\mathbf{u} \in E$). It is found that the theory extends without difficulties as long as we have *compact* operators. For a general bounded operator it may happen that the quantity $(T^{n*}T^n)^{1/2n}$ does not tend to a limit when $n \to \infty$. In the case of a compact operator in a Hilbert space, the spectrum of $T^{n*}(x)T^n(x)$ is discrete, the eigenvalues have finite multiplicities and they accumulate only at zero. In this case the characteristic exponents form a sequence tending to $-\infty$, and sometimes only finitely many of them are finite. A natural extension is possible also to *quasicompact* operators, where one has an essential spectrum with a certain diameter and, outside it, simple eigenvalues of finite multiplicities, which actually can accumulate on the essential spectrum.

Here follow two examples (Example 3 and section 9.2) which clarify somewhat how the multiplicative ergodic theorem applies to different theoretical contexts.

Example 3

Suppose we have six matrices T_i, $i = 1, \ldots, 6$, which, for instances, are 2×2 matrices of the form:

$$T_i = \begin{pmatrix} a_i & b_i \\ c_i & d_i \end{pmatrix}. \tag{62}$$

Then, suppose we perform an experiment simply by rolling a die, say, once each minute, and that this operation will continue forever. The state space Ω of the experiment is the set of infinite sequences $\omega = (\omega_1, \omega_2, \omega_3 \ldots)$ with $\omega_k \in V = \{1, \ldots, 6\}$ for each k. If we assign a probability $p_i = \frac{1}{6}$ to each possible outcome $i \in V$, a Bernoulli probability measure ρ on Ω is defined as we have seen in Example 2.

Next, we can get a dynamical system from this experiment by equipping the space Ω with a shift transformation $\sigma: \omega \to \omega' = (\omega_2, \omega_3, \ldots)$. Finally, we put ourselves in the framework of the multiplicative ergodic theorem by multiplying the matrices T_i depending on the outcomes of the experiment.

Let $\phi: \Omega \to 2 \times 2$ matrices be a map such that $\phi(\omega) = T_i$ if the first digit of ω is i. Then from the ordinary ergodic theorem we know that

$$\lim_{n \to \infty} \frac{1}{n} \sum_{k=0}^{n-1} \phi(\sigma^k \omega) = \lim_{n \to \infty} \frac{1}{n} \sum_{i=1}^{n} T_{\omega_i} = \frac{T_1 + T_2 + \cdots + T_6}{6}, \quad (63)$$

which is just the law of large numbers.

However, it is interesting to inquire whether there exist well defined properties governing the asymptotic behavior of the (random) products $T^n = \prod_{k=0}^{n-1} \phi(\sigma^k \omega) = T_{\omega_n} \ldots T_{\omega_1}$. This is just what the multiplicative ergodic theorem is about. In particular it says that the limit

$$\lim_{n \to \infty} \frac{1}{n} (T^{n*} T^n)^{1/2n}$$

does exist ρ-almost everywhere, providing a matrix Λ with well defined characteristic exponents – even though it gives no explicit expression for them.

An important remark is that the characteristic exponents depend in a noncontinuous manner on the matrix elements (a_i, b_i, c_i, d_i).

We shall see that this situation, namely the lack of an explicit expression for characteristic exponents and the lack of their continuity with respect to the parameters involved in the problem, is very general and raises some interesting problems about the interpretation of characteristic exponents as physical quantities.

Coming back to our example, let us point out that the theory of random products has been developed in particular by Furstenberg (1963), and, recently, thanks to the contribution of the Oseledec theorem, has found some interesting physical applications. One of them is in the theory of *localisation*.

9.2 Characteristic exponents in localisation theory

Here the problem is essentially to characterise the properties of the eigenfunctions of a one-dimensional Schrödinger equation in a random potential:

$$\psi_{i+1} - \psi_i + \psi_{i-1} = (E - V_i)\psi_i \tag{64}$$

in such a way as to satisfy some boundary conditions (for instance periodic). In solving this problem one gets in general either oscillating eigenfunctions or, and this is the interesting case, *localised* eigenfunctions.

If, for a given $\{V_i\}$, we find eigenfunctions which tend to zero exponentially when $i \to \pm\infty$ and for some value of E, then we have:

$$\lim_{N \to \infty} \frac{1}{N} \log \frac{|\psi_0|}{|\psi_{-N}|} = \lambda. \tag{65}$$

As we shall see λ can be interpreted as a characteristic exponent.

The relation with random matrix products emerges at once if we rewrite (64) in the form:

$$\begin{pmatrix} \psi_{i+1} \\ \psi_i \end{pmatrix} = \begin{pmatrix} E - V_i + 2 & -1 \\ 1 & 0 \end{pmatrix} \begin{pmatrix} \psi_i \\ \psi_{i-1} \end{pmatrix}. \tag{66}$$

Introducing the vector

$$\boldsymbol{\psi}_i = \begin{pmatrix} \psi_{i+1} \\ \psi_i \end{pmatrix},$$

(66) can be written in a more compact form, which shows that to solve a one-dimensional Schrödinger equation like (64) is actually the same thing as to iterate a map

$$\boldsymbol{\psi}_i = Q_i \boldsymbol{\psi}_{i-1}. \tag{67}$$

It is easy to see that the matrix Q_i is symplectic and therefore $\det Q_i = 1$. Equation (67) defines a process, and the properties of ψ_i are related to those of the infinite product $T^n = \prod_{i=1}^n Q_i$, with $n \to \infty$. Then, once again, the Oseledec theorem guarantees the existence of the limit $\lim_{n \to \infty} \frac{1}{n}(T^{n^*}T^n)^{1/2n} = \Lambda$. The matrix Λ can be cast in diagonal form:

$$\Lambda_d = \begin{pmatrix} \exp(\lambda_1) & 0 \\ 0 & \exp(\lambda_2) \end{pmatrix}, \tag{68}$$

where λ_1, λ_2 are the characteristic exponents. In this case, they satisfy the relation $\lambda_2 = -\lambda_1$ (because $\det \Lambda = 1$), and then the whole information on the decay properties of the wave function is contained in λ_1.

If we verify that $\lambda_1(E_n) > 0$ for some E_n, then – at least heuristically – we see that the corresponding eigenfunctions are exponentially localised. For more details on this subject we refer to Anderson (1968), Thouless (1972), and Kotani (1987).

9.3   Characteristic exponents in differentiable dynamics

We consider the case where $M = R^m$ and $\mathbf{f}$ is a differentiable map preserving an ergodic probability measure ρ, such that we have the time evolution

$$\mathbf{x}(n+1) = \mathbf{f}(\mathbf{x}(n)). \tag{69}$$

We denote by $T(x)$ the matrix $D_x f = (\partial f_i / \partial x_j)$ of partial derivatives of the components f_i at $\mathbf{x}$. Then we have

$$T_x^n = (D_{f^{n-1}x}f)(D_{f^{n-2}x}f) \ldots (D_{fx}f)D_x f = D_x f^n \tag{70}$$

where

$$f^n = \overbrace{f \circ \ldots \circ f}^{n \text{ times}}$$

Now, all the conditions of the multiplicative ergodic theorem are satisfied and the characteristic exponents $\lambda_1 \geq \lambda_2 \ldots$ are well defined. The finite set $\{\lambda_i\}$ captures the asymptotic behavior of the derivative along almost every orbit, in such a way that positive λ_i indicates eventual expansion and hence sensitive dependence on initial conditions, while those which are negative indicate contraction along certain directions.

As much as the exponential rate of growth of distances is given in general by λ_1 (i.e. if one picks out a vector at random then its growth rate is λ_1), the rate of growth of a k-directional Euclidean volume element is given by the sum $\lambda_1 + \lambda_2 + \ldots + \lambda_k$.

In particular the rate of growth of the m-directional volume element is the rate of growth of the jacobian $|J| = |\det(\partial f_i / \partial x_j)|$ and is given by $\lambda_1 + \ldots + \lambda_m$. This makes it possible to compute the lowest characteristic exponent once the larger ones and the jacobian of the transformation are known.

Consider, for instance, the Hénon mapping (17). We have already seen that it has a constant jacobian – see (18). Therefore the rate of growth of a surface element is given by $\lambda_1 + \lambda_2 = \log|J|$. Since $J = b = 0.3$ and numerically it is known that $\lambda_1 \approx 0.42$, we have that $\lambda_2 = \log|J| - \lambda_1 \approx -1.62$.

As another example, for a two-dimensional area preserving transformation we have $|J| = 1$, hence $\lambda_1 = -\lambda_2$.

The continuous case

The multiplicative ergodic theorem can be generalised without difficulties to the case of continuous-time dynamical systems – see, for example, (24). As a matter of fact, if we have a continuous time t we can write the linear operator $D_x f^t$ as a product of two linear operator $D_x f^t = (D_{f^n x} f^{t-n})(D_x f^n)$, where the first factor of the right-hand side is actually a bounded operator which does not change the asymptotic result (59).

We define the characteristic exponents for a continuous-time dynamical system (i.e. a *flow*) f^t as those for the discrete-time dynamical system generated by the time-one map $f = f^1$. This is a very general fact. If we discretise a continuous-time dynamical system by taking the time-one map, then all the interesting ergodic quantities (characteristic exponents, entropy, dimensions . . .) are unchanged – so that in the ergodic theory of differentiable dynamical systems there is no essential difference between discrete-time and continuous-time systems.

Now, if we have a flow in an m-dimensional space and we take a Poincaré section Σ, then we can obtain a discrete-time dynamical system in the $(m-1)$-dimensional surface of section by considering the first-return map P (see Fig. 16).

Of course, the original system has one characteristic exponent more than the derived one, but it is easy to realise that it vanishes. This corresponds to the fact that, on average, a vector carried along the direction of motion is neither increased nor contracted. Notice that this fact gives a further support to the possibility of studying first-return maps without loss of meaningful informations about the 'true' dynamical system, as we have discussed in chapter 3. Moreover we can relate the characteristic exponents λ_i of the continuous-time system (those which do not correspond to the direction of the motion) with the characteristic exponents $\tilde{\lambda}_i$ of the first-return map P:

$$\lambda_i = \frac{\tilde{\lambda}_i}{\langle \tau \rangle_\sigma}, \tag{71}$$

where $\langle \tau \rangle_\sigma$ is the average time between two crossings of Σ, computed with respect to the measure σ on Σ, that is the density of intersections of orbits with Σ.

Dynamical systems living on a manifold M

We consider now the case in which M is an m-dimensional manifold, say compact. In this case Df operates on the tangent bundle TM and we can use a Riemann metric on M to define characteristic exponents in terms of this metric. Alternatively, we can apply the multiplicative ergodic theorem after a decomposition of the manifold into cells. Namely, we need a partition of M into a certain number of pieces which are diffeomorphic to pieces of R^m, so that we can use Euclidean coordinates on each of them (see Fig. 17).

This operation introduces some discontinuities at the cuts, both in the map f and in the matrix T, because the correspondence between the manifold M and the different pieces which are homeomorphic to pieces of Euclidean space is not a continuous parametrisation. Nevertheless this does not matter at all because the multiplicative

Fig. 16. The Poincaré return map. From Guckenheimer and Holmes (1983).

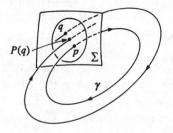

ergodic theorem requires functions only to be *measurable*. Furthermore different parametrisations will not get different results. In fact, even though the length of a vector depends on the parametrisation used, the ratio of the lengths corresponding to two different parametrisations of M is bounded and, therefore, the result obtained by taking the limit (61) is always the same.

Dynamical systems living in a Hilbert space E

Finally we consider the case of a flow f^t in an infinite-dimensional space, say a real Hilbert space E, where $D_x f^t$ is a compact linear operator for $t > 0$. For instance, the Navier–Stokes equation in two dimensions satisfies this compactness requirement, and in this case one can apply the multiplicative ergodic theorem. For the three-dimensional case the situation is more delicate because one has to avoid the singularities of the velocity field. More precisely, consider the Navier–Stokes equation (10) with $d = 3$, with a particular force $\vec{f}$ and suppose that the time evolution is well defined in a neighborhood of a compact set A which carries an invariant measure ρ. One can then restrict oneself to this compact set A and apply the Oseledec theorem.

Fig. 17. (a) The two-dimensional sphere S^2 cut into two pieces, each of them being diffeomorphic to a disk in R^2 by stereographic projection. (b) The two-dimensional torus T^2 cut into four rectangular pieces.

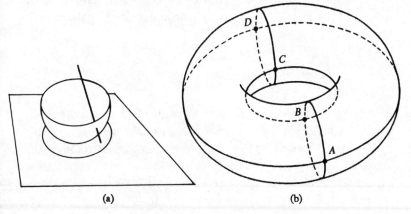

(a)   (b)

9.4   The spectrum of characteristic exponents as a classification tool

So far we have encountered a number of qualitatively different attractors, each of them being associated with a different type of time evolution. We now point out that the spectrum of characteristic exponents turns out to be a good indicator to deduce which kind of state the ergodic measure ρ is describing. To summarise, we have the following situations:

(1) Attracting fixed point

A *steady state* of a time evolution is associated with a point $Q \in M$ such that $f^t Q = Q$ for all t. There is a neighborhood U of Q so that $\lim_{t \to \infty} f^t x = Q$ for all $x \in U$. Clearly the measure $\rho = \delta_Q$ (where δ_Q is the Dirac's delta function at Q) is an invariant measure and, of course, ergodic. Since a volume element in M is contracted in all directions, the characteristic exponents are all negative. Using a symbolic notation developed by Crutchfield, in a three-dimensional space the spectrum of characteristic exponents would be written as $(-,-,-)$, see Fig. 18.

(2) Attracting limit cycle

Here we have $f^\tau Q = Q$ for some positive τ and there is a neighborhood U of the closed orbit $\Gamma = \{ f^t Q : t \in [0,\tau) \}$ so that $f^t x \to \Gamma$ as $t \to \infty$, for all $x \in U$. This situation is then described by the ergodic measure

$$\rho = \delta_\Gamma = \frac{1}{T} \int\limits_0^T dt \, \delta_{f^t Q},$$

namely ρ spread around evenly on Γ according to the time parameter. This attractor is naturally associated with a *periodic state* of a time evolution. In this case a volume element is not contracted along the direction of the orbit, thus the measure ρ has one characteristic exponents equal to zero and the others negative. Hence, in the three-dimensional case we obtain the spectrum $(0,-,-)$.

(3) Quasiperiodic attractor

In chapter 1 we discussed this case in great generality. Now, consider for simplicity the two-dimensional case, where the time evolution lies asymptotically on an attracting invariant torus T^2. This motion is associated with a *quasiperiodic state* with two frequencies ω_1, ω_2. We can represent it by a linear flow on T^2, namely by a translation in terms of suitable angular variables ϕ_1, ϕ_2:

$$\phi_i(t) = \phi_i(0) + \omega_i t \pmod{2\pi}, \; i = 1, 2.$$

If either $\omega_2 = 0$ or ω_1/ω_2 is rational, then we get periodic orbits. If ω_1/ω_2 is irrational then each orbit covers densely the torus as the time goes on. In this case one can prove that the flow f^t is uniquely ergodic,

Fig. 18. From Shaw (1981).

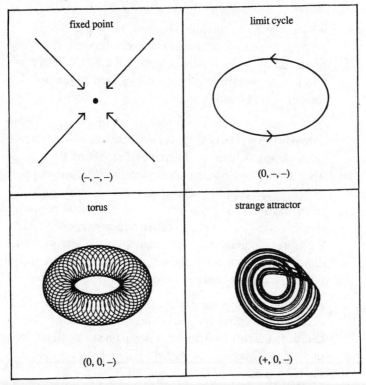

i.e. the invariant ergodic measure ρ is the same for all initial conditions, where ρ is the Haar measure $(2\pi)^{-2}d\phi_1 d\phi_2$. The spectrum of characteristic exponents is $(0,0,-)$, as Fig. 18 shows. In the general case of k frequencies we have that k characteristic exponents vanish and the others are negative.

(4) Strange attractor

This is the case which is associated with a *chaotic state* of the time evolution. It is described by the *asymptotic measure* ρ carried by the attractor – see chapter 5. If any one of the characteristic exponents is positive a volume element is expanded in some direction at exponential rate and neighboring trajectories are diverging. In Fig. 18 the Rössler attractor is plotted, with one positive characteristic exponent. Four-dimensional chaotic systems may be classified into type $(+,0,-,-)$ and $(+,+,0,-)$.

Finally, we want to make the following remarks:

For a discrete-time dynamical system, with $t \geq 0$, chaos may occur in *one* (or more) dimension(s). We have already seen this for the quadratic (noninvertible) map $x \rightarrow \mu x(1-x)$ of the unit interval into itself.

In the case of a diffeomorphism, i.e. for a discrete-time dynamical system where t may be negative as well as positive, chaos occurs only in two (or more) dimensions. (Indeed, we want at least one positive characteristic exponent. Suppose that ρ is an ergodic measure for a one-dimensional diffeomorphism and has a positive characteristic exponent, then one can show that ρ is carried by a periodic orbit and is therefore nonchaotic.) An example of chaotic two-dimensional attractor is provided by the Hénon mapping (although a mathematical proof of this is still missing).

For a continuous-time dynamical system (i.e. a flow) chaos occurs only in three (or more) dimensions. (The situation is similar to that of diffeomorphisms, but one extra dimension – in the flow direction – is needed.) Examples are provided by the Lorenz attractor (Fig. 5) and the Rössler attractor (Fig. 10).

9.5   Parameter dependence

If the dynamical system depends continuously on a bifurcation parameter μ, it is interesting to inquire how the ergodic invariant measure ρ and its spectrum of characteristic exponents change as μ is varied. In particular one would like to know whether there are some continuity properties of the characteristic exponents λ_i as functions of μ. This problem appears to be quite difficult and is not yet well understood.

There are some simple cases, like cases (1) and (2) discussed in section 9.4, where the λ_i's depend continuously on μ, but these situations are rather exceptional. Already for the quasiperiodic case the situation gets more delicate. For $k \geq 2$, for instance, if for $\mu = \mu_0$ there is a quasiperiodic attractor, it may happen that for μ close to μ_0 frequency locking leads to the rational case, with negative characteristic exponents. For $k \geq 3$ (see chapter 2) it is likely that, for μ arbitrarily close to μ_0, the system possesses a strange attractor with some positive characteristic exponents.

In general cases, for each μ there may be several attractors each having at least one physical measure ρ_k^μ and, because of complicated mechanisms of captures and explosions, the dependence of these measures ρ_k^μ on μ may not be continuous. Furthermore, even though ρ_k^μ depends continuously on μ, it may be not the same for the characteristic exponents λ_i^μ. In general, *characteristic exponents are discontinuous functions of the bifurcation parameter μ.*

This discontinuous behavior raises some problems about the interpretation of characteristic exponents as physical quantities. In fact, since a physical quantity (i.e. a measurable quantity) has a role in predictions, should it not have at least a piecewise continuous dependence on the parameters which control the physical state of the system? In other words, since characteristic exponents *measure* the amount of dynamical instability in a system, should they not have some stability under small changes of the external parameters? As long as dynamical *instability* is the object of measurement and prediction, it should be a *structurally* stable property; just in order to make measurements and predictions legitimate.

Actually the resolution of this problem lies largely in the fact that physical measurements are always smoothed by the instrumental procedure, and we have to consider as meaningful just what this procedure brings forth.

We give now an example to illustrate how this problem is shaped in computer studies of dynamical systems.

Example 4

Consider again the map $f_\mu(x) = \mu x(1 - x)$ of the unit interval $[0,1]$ into itself, where the parameter μ ranges over the interval $[0,4]$. We have seen that in the parameter region $[0,3.57\ldots)$ this map exhibits simple behavior (fixed points or periodic orbits). For $\mu_\infty = 3.57\ldots$ there is the Feigenbaum attractor (see Fig. 9) with the characteristic exponent $\lambda = 0$. Then we have considered the case $\mu_{\max} = 4$ where the time evolution is described asymptotically by the ergodic measure (49), which is absolutely continuous with respect to Lebesgue measure m and has $\lambda = \log 2$. Let us consider now the parameter region $(\mu_\infty, \mu_{\max})$, namely the so-called *aperiodic* region.

Even here there are several *windows* (i.e. open intervals) where the map f_μ has stable periodic orbits. In general windows tend to get shorter as the periods get longer – but nevertheless they constitute an infinite set where the characteristic exponent λ is negative. Furthermore one is still far from a complete understanding of the ergodic properties of f_μ in the parameter region obtained by discarding all windows. It is believed that the set $\{\mu : \lambda > 0\}$ has a positive Lebesgue measure and, in any case, this is so for the set of parameter values where f_μ has an absolutely continuous invariant measure (see Collet and Eckmann, 1980a, 1983); Jakobson, 1981. The latter situation can occur, for instance, when the image of the critical point $\frac{1}{2}$ (the maximum) under some iterate of f_μ falls onto an unstable fixed point or an unstable periodic orbit.

A nice computer experiment performed first by Shaw (1981) and then improved by Crutchfield, shows the dependence of the characteristic exponent λ on the parameter μ. What has actually been calculated is the quantity

$$\lambda_n(\mu) = \frac{1}{n} \sum_{i=0}^{n-1} \log |f'_\mu(f^i_\mu(0))|$$

for a certain number N of values of μ, spaced δ apart. Fig. 19a shows the curve obtained by taking $N = 300$ points, with $\delta = 0.002$, each representing 100 000 iterations of f_μ.

The curve $\mu \to \lambda(\mu)$ is roughly increasing as μ increases, reaching finally the value log 2 when the map becomes strictly two-to-one. Yet its shape reveals a nonsmooth, actually very complicated, behavior of $\lambda(\mu)$. It might be argued that since λ is an asymptotic quantity, involving the limit $n \to \infty$, the great complexity of the curve appears progressively as n gets larger and larger.

Now, a natural question arises about the meaning of what one is measuring. Suppose that with a given grid spacing (i.e. a given δ) it is found, for example, that $\lambda(\mu)$ is positive in the interval $[\mu_1, \mu_2]$. Then it is likely that, if one improves the precision of his procedure, this will introduce a smaller interval $[\mu_3, \mu_4] \subset [\mu_1, \mu_2]$ where $\lambda(\mu)$ is negative. This is just what the lower curve in Fig. 19b (which is obtained by reducing the grid spacing δ for the dashed region of Fig. 19a) shows. Hence, we see that what we can consider as a meaningful result depends strongly on our measurement procedure; but nevertheless this does not diminish at all the possibility to extract physical sense from these experiments.

The upper curve in Fig. 19b represents the *topological entropy* $h(f_\mu)$ of the map f_μ. There are several possible definitions of $h(f_\mu)$ (see later) but in this case we can define it simply by

$$h(f_\mu) = \lim_{n \to \infty} \frac{1}{n} \log N_n, \tag{72}$$

where N_n is the number of *laps* (i.e. monotone pieces) of $f_\mu^n (= 1 +$ the number of critical points of f_μ^n). This is an upper bound for the characteristic exponent λ (see Ruelle, 1978), and, on the contrary of the latter, it appears to be a continuous function of the parameter μ.

To conclude, let us mention that a similar computer experiment, computing λ_1 as a function of the parameter a for the Hénon transformation (17) (with $b = 0.3$), has been carried out by Feit (1978). The result, shown in Fig. 20, strongly resembles Fig. 19, but it is even less well understood.

Fig. 19. Characteristic exponent for the quadratic map $x \to \mu x(1-x)$.

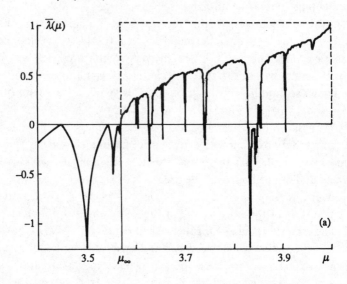

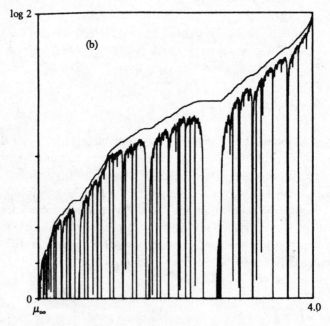

9.6 Experimental determination of characteristic exponents

We know that characteristic exponents are computed from the *derivative* $D_x f^t$. Since in computer experiments the derivative is often directly calculable, whereas in physical experiments it has to be estimated by an analysis of the experimental data, the methods are somewhat different in the two cases. The main difference lies in the fact that, when dealing with experimental data, the calculation of the characteristic exponents needs some preliminary steps in order to reconstruct the dynamics in a suitable space. This is a very general procedure: whenever we want to extract some useful information from a time series of the form $\{x_i\} = \{x(i\tau)\}$, $1 \leq i \leq N$, obtained by monitoring a scalar signal for a finite time $T = N\tau$ and with finite precision, the first step is to embed it in a state space of finite dimension d_E (the *embedding dimension*), and to reconstruct a d_E-dimensional orbit. For this purpose one may follow, for instance, the time delays method (see chapter 6).

Then one tries to obtain the tangent maps to this reconstructed dynamics, for instance by a least-squares fit (see Eckmann *et al.*, 1986), and, finally, one deduces the characteristic exponents from the tangent maps. This latter step constitutes an analogous problem both for computer experiments and physical experiments.

Fig. 20. Characteristic exponent for the Hénon map.

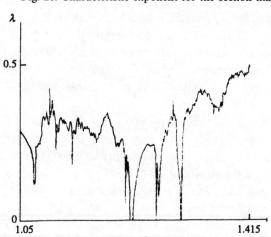

10

Invariant manifolds

We find it convenient to introduce *stable* and *unstable* *manifolds* starting with the two-dimensional map we introduced in chapter 3, the Hénon map. Consider the diffeomorphic map of the plane to itself:

$$f(x_1,x_2) = (x_2 + 1 - ax_1^2, bx_1) \tag{73}$$

f is invertible with inverse

$$f^{-1}(x_1,x_2) = (b^{-1}x_2, x_1 - 1 + ab^{-2}x_2^2). \tag{74}$$

Even though there is no mathematical proof of this fact, we have seen numerically that, when $a = 1.4$ and $b = 0.3$, the map (73) has two characteristic exponents λ_1, λ_2 such that $\lambda_2 < 0 < \lambda_1$. Then, the multiplicative ergodic theorem asserts the existence of two *linear* spaces $E_x^{(1)} \supset E_x^{(2)}$ (actually $E_x^{(1)}$ coincides with R^2) such that

$$\lim_{n \to \infty} \frac{1}{n} \log \|D_x f^n \mathbf{u}\| = \lambda_1$$

if $\mathbf{u} \in E_x^{(2)} \backslash E_x^{(2)}$. The rate of growth of $\mathbf{u}$ is $\lambda_1 > 0$ and hence $\mathbf{u}$ is expanded by the iterates of Df. Conversely, if $\mathbf{u}$ lies in the subspace $E_x^{(2)}$ then its rate of growth is $\lambda_2 < 0$ and it will be contracted.

The direction $E_x^{(2)}$ where vectors are contracted is called the *stable direction* and it is also denoted by E_x^s. Similarly, we denote by E_x^u the *unstable direction*, which consists of vectors contracted by Df^{-1} (see Fig. 21).

Now we can ask to what extent properties of iterates of Df are reflected in properties of iterates of f itself. An important property of f is the existence of invariant *stable* and *unstable manifolds* which are the nonlinear analogues of E_x^s and E_x^u for Df.

The stable manifold, denoted by V_x^s, is a smooth invariant curve (in the general case it is a manifold), with dimension equal to that of E_x^s, passing through x tangent to E_x^s, and composed of points y such that

$d(f^n x, f^n y) \to 0$ when $n \to \infty$, exponentially like $\exp(n\lambda_2)$. Here $d(x,y)$ is the Euclidean distance of x and y, but in different situations d may be Riemann distance or the norm distance in a Hilbert space.

In general, whenever the multiplicative ergodic theorem asserts that there is a linear space $E^{(i)s}$ where exponential contraction takes place, we can get a (unique) nonlinear version $V^{(i)s}$ of it, where contraction occurs at the same exponential rate λ_i. These *global* manifolds have the feature that they tend to fold and accumulate in a very messy manner, as suggested by Fig. 22.

We can also define a *local* stable manifold. Let $\varepsilon > 0$, and $\lambda_2 < \lambda < 0$, then we write

$$V^s_x(\lambda, \varepsilon) = \{y : d(f^n x, f^n y) \leq \varepsilon \exp(\lambda n) \text{ for all } n \geq 0\}. \tag{75}$$

And V^s_x can be written in terms of $V^s_x(\lambda, \varepsilon)$ as

$$V^s_x = \bigcup_{n > 0} f^{-n} V^s_x(\lambda, \varepsilon). \tag{76}$$

Fig. 21. Stable and unstable directions for the Hénon attractor. From Grassberger (1984).

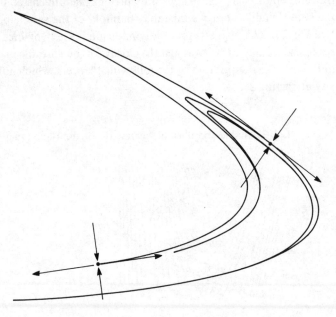

The *local* unstable manifold $V_x^u(\lambda,\varepsilon)$, as well as the full unstable manifold V_x^u, are defined in the same way, replacing n with $-n$ in definitions. Thus, if $0 < \lambda < \lambda_1$, we have

$$V_x^u(\lambda,\varepsilon) = \{y: d(f^{-n}x, f^{-n}y) \le \varepsilon\exp(\lambda n) \text{ for all } n \ge 0\} \quad (77)$$

and

$$V_x^u = \bigcup_{n>0} f^n V_x^u(\lambda,\varepsilon). \quad (78)$$

For more general dynamical systems all these definitions extend in a straightforward way (see Eckmann and Ruelle, 1985).

We have seen that the attractor generated by iterating the Hénon transformation has the appearence of a line folded over many times (see Fig. 7b). Moreover it is immediate to verify that f has two fixed points whose coordinates are given by

$$x_1 = \frac{(b-1) \pm [(1-b)^2 + 4a]^{\frac{1}{2}}}{2a}, x_2 = bx_1.$$

For the parameter values $a = 1.4$ and $b = 0.3$ the fixed point P in the first quadrant is of *saddle type*, and it is found that the points of the trajectory plotted in Fig. 7b appear to fill out a one-dimensional curve indistinguishable from the unstable manifold of the saddle point P (see Fig. 23). This suggests that the complicated system of lines which forms the Hénon attractor is just the closure of the unstable manifold of P. This is also supported by the following theorem, which holds for any attracting set A.

Fig. 22. Stable and unstable manifolds for the Hénon map.

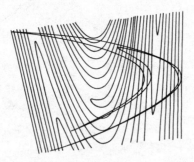

Theorem

If A is an attracting set, and $x \in A$, then $V_x^u \subset A$, i.e. the unstable manifold of x is contained in A.

This result has the natural consequence that the number of positive characteristic exponents is a lower bound to the dimension of A, and therefore it constitutes the smallest number of 'effective' degrees of freedom of a chaotic motion, as we have already mentioned in chapter 1. Another fact which tends to confirm the previous statement is that in many cases the physical measure ρ carried by an attractor is *absolutely continuous on unstable manifolds*, as we shall see later.

To summarise, the local structure of the attractor visible in Fig. 7b (as well as for other two-dimensional time evolutions generated by a diffeomorphism) turns out to be a manifold (line) in the unstable direction (note however the folds where the stable and unstable manifolds are tangent). The measures carried by the attractor are singular with respect to Lebesgue measure (they have very 'rough' density, or more exactly density cannot be defined).

Fig. 23. A saddle point. From Smale (1967).

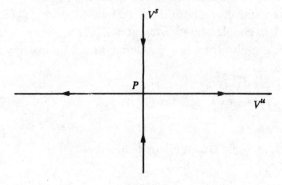

11

Axiom A and structural stability

It is interesting now to inquire into the conditions under which the asymptotic behavior of a dynamical system is stable under small perturbations of the system itself, i.e. is *structurally stable*. This is particularly important if we require that this behavior corresponds to verifiable physical properties of the system.

For this purpose we need some definitions.

Let $f: M \to M$ be a diffeomorphism of a compact manifold. We say that a point $x \in M$ is *wandering* if it has a neighborhood U so that $f^n(U) \cap U = \emptyset$ for n large enough. The *nonwandering* set $\Omega(f)$ is the set of all points which are not wandering (for instance, the periodic points of f are clearly nonwandering), see Fig. 24.

$\Omega(f)$ is a compact f-invariant subset of M, and contains the attractors of the dynamical system.

Next, we say that a compact set $A \subset M$ is *hyperbolic* (for f) if there exists a direct sum decomposition of the tangent space $T_x M$ at each point $x \in A$ into conplementary subspaces E_x^s, E_x^u depending continuously on x, and also constants $K > 0$ and $0 < \lambda < 1$, satisfying for all $n \geq 0$:

$$
\begin{aligned}
(D_x f) E_x^s &= E_{fx}^s \\
(D_x f) E_x^u &= E_{fx}^u
\end{aligned}
\tag{79}
$$

Fig. 24. From Guckenheimer and Holmes (1983).

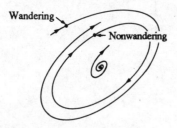

and

$$\|D_x f^n(\mathbf{u})\| \le K\lambda^n \|\mathbf{u}\| \text{ if } \mathbf{u} \in E_x^s$$
$$\|D_x f^{-n}(\mathbf{u})\| \le K\lambda^n \|\mathbf{u}\| \text{ if } \mathbf{u} \in E_x^u, \tag{80}$$

An analogous definition holds for continuous flows, provided one allows one 'extra' subspace E_x^0 in the flow direction, which is neither contracting nor expanding.

We say that f satisfies *Axiom A* if

(i) $\Omega(f)$ is hyperbolic, and

(ii) the periodic points of f are dense in $\Omega(f)$.

Axiom A dynamical systems were introduced by Smale in 1967. In particular they include *Anosov* diffeomorphisms, for which the whole manifold M is hyperbolic. As we have seen in chapter 1, this class of dynamical systems originated with the work of Hadamard, when he studied the geodesic flow on the unit tangent bundle of a manifold with constant negative curvature: this is what we call today an Anosov flow. Other examples are *gradient systems*, Smale's *horseshoe* and the *solenoid* (for details see Smale, 1967; Bowen, 1978).

We introduce now the concept of *structural stability*. First let us say that two diffeomorphisms $f,g: M \to M$ are *topologically equivalent* if there exists a homeomorphism h such that $h \circ f = g \circ h$. This definition implies that h takes an orbit $\{f^n(x_0)\}$ to an orbit $\{g^n(x_0)\}$.

Then, a diffeomorphism f of a compact manifold M is structurally stable if it has a neighborhood V in the C^1 topology (i.e. f and g are close if both the diffeomorphisms and their (partial) derivatives are close), such that every g in V is topologically equivalent to f (to be specific, this is C^1 structural stability).

The concept of structural stability, first introduced by Andronov and Pontrjagin in 1937, is completely different from that of dynamical stability (or Liapunov stability). The latter refers to individual orbits and requires that there is no sensitive dependence on initial conditions. The former refers to the whole system and asks that, under small C^1 perturbations of the system, the qualitative features of the entire time evolution are preserved (see Thom, 1975). It has been known for some time that if f satisfies Axiom A together with a hypothesis of *strong transversality*, then it is structurally stable.

By strong transversality we mean that every intersection point of

V_x^s and V_y^u is transversal for all $x,y \in \Omega(f)$ (V_x^s and V_y^u intersect transversally in a point $z = V_x^s \cap V_y^u$, if the tangent spaces $T_z V_x^s, T_z V_y^u$ span $T_x M$, see Fig. 25).

A recent result by Mañé provides necessary conditions for structural stability.

Theorem

A diffeomorphism $f: M \to M$ is C^1 structurally stable if and only if it satisfies Axiom A and strong transversality.

In the case of the Hénon diffeomorphism we have seen that the unstable manifolds fold back upon themselves and have tangencies with the stable manifolds: the Hénon attractor cannot therefore be structurally stable.

Fig. 25. Transversal and nontransversal intersections of curves in $\mathbf{R}^2$. From Janich (1984).

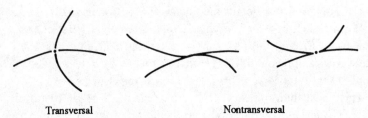

Transversal   Nontransversal

12

Entropy

From previous discussions we know that there exist deterministic dynamical systems where trajectories emerging from nearby initial conditions diverge exponentially. Due to this sensitivity any uncertainty about seemingly insignificant digits in the sequence of numbers which defines an initial condition, spreads with time towards the significant digits, leading to chaotic behavior. Therefore there is a change in the *information* we have about the state of the system. This change can be thought of as a *creation* of information if we consider that two initial conditions that are different but indistinguishable (within a certain precision), evolve into distinguishable states after a finite time.

If f is a transformation preserving a measure ρ, then the *Kolmogorov–Sinai invariant*, or entropy, denoted by $h(\rho)$, measures the asymptotic rate of creation of information by iterating f.

This concept was introduced by Kolmogorov in 1958, when looking for a number which was invariant under isomorphisms. The problem of deciding when two measure-preserving transformations are equivalent, or isomorphic, is one of the fundamental problems of abstract ergodic theory. For instance, the problem of whether the two Bernoulli shifts $\{\frac{1}{2},\frac{1}{2}\}$ and $\{\frac{1}{3},\frac{1}{3},\frac{1}{3}\}$ are equivalent obtained no solution for many years, until it was shown that they have different entropies and hence they are nonisomorphic.

A notion of entropy (which can be applied to shifts) was first introduced by Shannon in 1948, in a famous work which originated information theory. Suppose we perform an experiment with m possible outcomes, for example rolling a die with m faces. Let $p_1, p_2, \ldots, p_m$ be the probabilities of the different outcomes. Then, a measure of the amount of uncertainty in the experiment – namely the amount of uncertainty about which outcome will turn out, *before* each observation – is given by the function

$$H(p_1,p_2,\ldots,p_m)=-\sum_{i=1}^{m}p_i\log p_i \qquad (81)$$

with the understanding that $u\log u = 0$ when $u=0$. From (81) it is clear that the two Bernoulli shifts mentioned above have different entropies, respectively $\log 2$ and $\log 3$. (Note, however, that it remains to show that the entropy is the same for isomorphic systems!) For a more general dynamical system (not a shift) the notion of entropy must be somewhat modified, introducing the Kolmogorov–Sinai invariant. We shall assume that the support of ρ is a compact set with a given metric. Let $\mathscr{R}=(\mathscr{R}_1,\mathscr{R}_2,\ldots,\mathscr{R}_m)$ be a finite partition of the support of ρ, such that the diameter of each $\mathscr{R}_i$ corresponds to the precision we are working with. In other words, given a certain precision, we construct a partition in such a way that, inside each piece of it, we cannot distinguish between two different points. Then, for each piece $\mathscr{R}_i$ we write $f^{-k}\mathscr{R}_i$ for the set of points mapped by f^k to $\mathscr{R}_i$, and we denote by $f^{-k}\mathscr{R}$ the partition $(f^{-k}\mathscr{R}_1,\ldots,f^{-k}\mathscr{R}_m)$. Next, we define the partition $\mathscr{R}^{(n)}$ as the partition whose pieces are the intersections

$$\mathscr{R}_{i_1}\cap f^{-1}\mathscr{R}_{i_2}\cap\ldots\cap f^{-n+1}\mathscr{R}_{i_n}$$

with $i_j\in\{1,2,\ldots,m\}$, and we write $\mathscr{R}^{(n)}$ as

$$\mathscr{R}^{(n)}=\mathscr{R}\vee f^{-1}\mathscr{R}\vee\ldots\vee f^{-n+1}\mathscr{R}.$$

One realises at once that, for increasing $n,\mathscr{R}^{(n)}$ is a new partition (generated by $\mathscr{R}$ in a time interval n) which is finer and finer.

Now, the information content of the partition $\mathscr{R}$ with respect to the measure ρ is given by

$$H(\mathscr{R})=-\sum_{i=1}^{m}\rho(\mathscr{R}_i)\log\rho(\mathscr{R}_i), \qquad (82)$$

and the limit

$$h(\rho,\mathscr{R})=\lim_{n\to\infty}[H(\mathscr{R}^{(n+1)})-H(\mathscr{R}^{(n)})]=\lim_{n\to\infty}\frac{1}{n}H(\mathscr{R}^{(n)}) \qquad (83)$$

exists and gives the rate of information creation with respect to the partition $\mathscr{R}$. Finally, the entropy $h(\rho)$ is defined as the limit for finer and finer partitions:

$$h(\rho)=\sup h(\mathscr{R},\rho), \qquad (84)$$

where the supremum extends over all finite partitions of the support of ρ. From this definition it is clear that isomorphic systems have the same entropy. There is a case where the supremum may be avoided in (84), i.e. $h(\mathcal{R},\rho)=h(\rho)$; this happens when $\mathcal{R}$ is a *generating partition* (see Billingsley, 1965), and this is how the entropy for Bernoulli shifts can be computed.

We want now to point out that the Shannon definition (81) of entropy (on which the Kolmogorov definition is based as well) is not the only possibility. Given a probability distribution $\{p_1,p_2,\ldots,p_m\}$, the most important generalised entropies are Renyi's informations of order α, defined as

$$H_\alpha(p_1,p_2,\ldots,p_m)=\frac{1}{1-\alpha}\log\sum_{i=1}^{m}p_i^\alpha, \tag{85}$$

where α is a positive real number not equal to one. The H_α's are indeed generalisations of the Shannon entropy $H(\lim_{\alpha\to 1}H_\alpha=H$ holds), and they provide upper bounds (for $\alpha<1$) and lower bounds (for $\alpha>1$) for it (see Grassberger, 1984).

Let us consider the case $\alpha=2$. We have $H_2(p_1,\ldots,p_m)=-\log\sum_{i=1}^{m}p_i^2$, that translated in the frame of the Kolmogorov definition writes

$$H_2(\mathcal{R})=-\log\sum_{i=1}^{m}\rho(\mathcal{R}_i)^2. \tag{86}$$

Then we define the entropy

$$K_2(\rho)=\sup_{\mathcal{R}^n}\lim_{n\to\infty}H_2(\mathcal{R}^{(n)}) \tag{87}$$

if the limit exists. The entropy $K_2(\rho)$ is interesting because it provides a lower bound to the entropy $h(\rho)$

$$K_2(\rho)\leq h(\rho)$$

and is more easily accessible in experimental situations (see Grassberger and Procaccia, 1983a,b).

12.1 Entropy and characteristic exponents

We turn now our attention to those results which relate entropy to characteristic exponents. Intuitively, since uncertainty is caused by exponential separation of nearby points, one should expect the entropy to be related to the *positive* characteristic exponents.

Consider an m-dimensional ball of radius ε. If we let this volume element evolve according to the transformation f, we see that it will be deformed into an ellipsoid (see Fig. 26).

In particular, for the semi-axes of the ellipsoid that are stretched by the transformation, we have:

$$\varepsilon_i \to \varepsilon_i \exp(\lambda_i n)$$

so that two indistinguishable points lying on an expanding axis (i.e. an unstable direction) become distinguishable after a finite time – thus providing creation of information. Now, if the invariant measure ρ spreads out uniformly over the ball, then the mean rate of creation of information due to the stretching is just λ_i. In the general case, considering any direction, we will deal with a product $\Pi_{i=1}^{k} \exp(n\lambda_i)$, where λ_k is the last positive characteristic exponent and we define the *unstable dimension* m_+ as the sum of the multiplicities of the positive characteristic exponents. We see that if the invariant measure ρ is absolutely continuous (for the definition of an absolutely continuous measure see Example 2, chapter 7), the rate of creation of information is just the mean rate of expansion of m_+-dimensional volume elements. However, if ρ is singular along unstable directions (the directions where we get information from the separation of nearby points) then we have still creation of information, but its rate is strictly less than the rate of expansion. To summarise, we have the following general results:

Theorem

(a) (Ruelle, 1978). Let $f: M \to M$ be a C^1 map of a compact manifold preserving an ergodic measure ρ. Then

Fig. 26. Ball deformed into an ellipsoid.

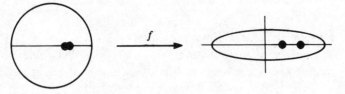

$$h(\rho) \le \sum_{\lambda_i > 0} \lambda_i. \tag{88}$$

(*b*) (Pesin, 1977). If f is a diffeomorphism with Hölder derivatives, and ρ is absolutely continuous with respect to the Lebesgue measure of M, then

$$h(\rho) = \sum_{\lambda_i > 0} \lambda_i. \tag{89}$$

More generally, the Pesin identity holds for SRB measures (from Sinai, Ruelle, Bowen), namely measures that are smooth along unstable directions.

12.2 SRB measures

We define an SRB measure in the following way. First, consider a ρ-measurable set of the form $S = \bigcup_{\alpha \in A} S_\alpha$, where the S_α are disjoint pieces of unstable manifolds V^u (each S_α can be constructed by the intersection of a *local* unstable manifold with S) (see Fig. 27).

Then we can disintegrate the invariant measure ρ along the elements S_α of S

$$\rho(\bullet) = \int \rho_\alpha(\bullet) m(d\alpha), \tag{90}$$

where m is a measure on A and the *conditional probability measures* ρ_α have support in the S_α's.

Next, we consider the case where the conditional measures ρ_α are absolutely continuous with respect to the Lebesgue measure on V^u. In this case we can write

$$\rho_\alpha(dx) = \phi_\alpha(x) dx \text{ on } S_\alpha, \tag{91}$$

Fig. 27. The set S is the union of disjoint pieces S_α.

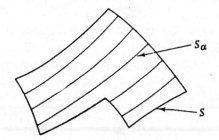

where dx denotes the volume element of S_α and ϕ_α is an integrable function (i.e. $\phi_\alpha \in L^1$).

Finally, we say that an ergodic measure ρ is SRB if its conditional probability measures ρ_α are absolutely continuous with respect to the Lebesgue measure for some choice of S with $\rho(S) > 0$. This definition is actually independent of the choice of S and its decomposition.

We have the following result:

Theorem (Ledrappier and Young, 1985)

Let f be a C^2 diffeomorphism of an m-dimensional manifold M, and ρ an ergodic measure with compact support. Then, the following conditions are equivalent:

(a) ρ is an SRB measure;

(b) ρ satisfies the Pesin identity (89).

Furthermore, these conditions imply that the density functions ϕ_α are C^1.

In chapter 8 we mentioned that SRB measures are good candidates to describe physical measures. In general, the selecting arguments for a physical measure are of two kinds. First we have mentioned stability under small stochastic perturbations: instead of a deterministic time evolution we consider a time evolution with some noise added to it, then we look for the stationary measure and finally we let the noise tend to zero. Let us see intuitively how this works. We think of an experiment corresponding to a time evolution where a volume element (an m-dimensional ball of radius ε) is stretched in some directions and contracted in some other directions (see Fig. 26), and where, at each step, we add some noise of amplitude ε. If we repeat these two operations (evolving by time evolution and putting some noise) again and again, then a heuristic argument indicates that we will get an invariant measure which is smooth along unstable directions. This is because the deterministic part of the time evolution will improve the continuity of the density in the unstable directions by stretching, and roughen it in other directions due to contraction. Thus (if all goes well) in the zero-noise limit $\varepsilon \to 0$ we will get a measure ρ that satisfies SRB conditions.

Another way of selecting the measure is by taking a point x_0 *at random with respect to the Lebesgue measure* on the manifold M (the Lebesgue measure is a more natural notion of sampling than the invariant measure ρ, because the latter is usually singular), and then take the time average (43). If ρ is an SRB measure with no zero characteristic exponent,

$$\frac{1}{n}\sum_{k=0}^{n-1}\delta_{f^k x_0}$$

tend to the SRB measure ρ when $n\to\infty$, not just for ρ-almost all x_0, but for x_0 in a set of positive Lebesgue measure.

Hence it turns out that in many cases these two methods give the same result. One important case is that of Axiom A systems.

Axiom A systems are the prototype with respect to which we try to obtain a general theory of differentiable dynamical systems; this is true in particular for the study of asymptotic invariant measures. If f is an Axiom A diffeomorphism on a compact manifold M, it is known that the nonwandering set $\Omega(f)$ (see chapter 11), is the union of finitely many closed sets $\Omega_1, \ldots, \Omega_s$, called *basic sets*. Each Ω_i is invariant (i.e. $f(\Omega_i)=\Omega_i$), and $f|\Omega_i$ is *transitive*, namely there exist $x\in\Omega_i$ such that $\{f^n x : n\geq 0\}$ is dense in Ω_i. Those sets Ω_i that are attracting sets are called *attractors*.

Let us now look at one of these attractors, with basin of attraction U. Then, there is a theorem (Ruelle, 1976) which asserts that if we exclude from U a set B with Lebesgue measure zero (the Lebesgue measure on the manifold M) then whenever $x\in U\backslash B$ the time average

$$\lim_{n\to\infty}\frac{1}{n}\sum_{k=0}^{n-1}\delta_{f^k x}$$

provides a unique SRB measure with support on the attractor.

An analogous result is obtained by considering stationary measures associated with stochastic processes. In general, a *stochastic dynamical system*, obtained from a map $f: M\to M$ by adding some noise, is a time evolution defined no longer on M, but at the level of probability measures on M (Eckmann and Ruelle, 1985). For Axiom A systems, if there are r attractors $A_1, \ldots, A_r$ and $\rho_1, \ldots, \rho_r$ are the corresponding SRB measures, then, given a stationary measure ρ_i^ε with support near A_i, it is known (Kifer, 1974) that $\rho_i^\varepsilon\to\rho_i$ when $\varepsilon\to 0$. Thus for an Axiom A system the notion of SRB measure and

Kolmogorov measure are essentially equivalent. Moreover, given a basic set Ω_i, it is possible to apply a statistical mechanics formalism using the device of *Markov partitions*, and, if Ω_i is an attractor, the SRB measure we find on it corresponds to a *Gibbs state* of this statistical mechanics (see Sinai, 1972; Bowen, 1975, 1978; Bowen and Ruelle, 1975). The uniqueness of such a measure on an Axiom A attractor corresponds to the uniqueness of the Gibbs state for a one-dimensional lattice spin system with fast decreasing interactions. In more general cases (non-Axiom A), usually a Markov partition of the attractor is not available and hence it is difficult to translate these concepts in a straightforward way. It is known that a one-dimensional lattice system with long range interactions can have several distinct Gibbs states (i.e. *phase transitions* may occur), and it would be interesting to see such a situation occurring for dynamical systems.

There are examples where no SRB measures exist and therefore the physical measure is not absolutely continuous with respect to the Lebesgue measure on the unstable manifolds.

Counterexample (Bowen, 1975)

Consider a flow in R^2 with three fixed points A, B, C, where A, C are repelling and B of saddle type (see Fig. 28).

It can be seen that if one starts anywhere else than at the points A, C then one gets a time average that converges to a Dirac δ at B. Clearly this measure is not absolutely continuous along the unstable manifold (the 'figure ∞'), and, even though the entropy is zero, the system possesses one strictly positive characteristic exponent (and another one strictly negative). So, the Pesin identity does not hold.

Fig. 28. Bowen's counterexample.

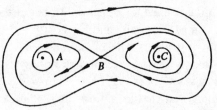

Notice that the measure δ_B is stable under small stochastic perturbations (i.e. it is a Kolomogorov measure). The heuristic argument that a measure stable under small perturbations is SRB is therefore invalid here.

We have already mentioned the following fact:

Theorem (Pugh and Shub, 1984)

Let f be a C^2 diffeomorphism on a compact manifold M and ρ an SRB measure such that all characteristic exponents are different from zero. Then there is a set $S \subset M$ with positive Lebesgue measure such that

$$\lim_{n \to \infty} \frac{1}{n} \sum_{k=0}^{n-1} \delta_{f^k x} = \rho$$

for all $x \in S$.

This theorem unfortunately provides no control of what happens if some noise is added to the system. In fact, since the set S is not the complement of a set of Lebesgue measure zero, in a neighborhood of the support of ρ we may think of it as having 'holes' so that, by adding a small noise, a point x_0, initially in S, may fall into one of these holes and the asymptotic behavior will be completely unknown.

It is believed that for the Hénon attractor there is a situation of this type, at least for certain values of a and b. In this case we may expect that, using a very high precision, one would see that in a neighborhood of the attractor there is a set (with a very complicated structure and probably very small in measure), such that any initial point in this set will end up in a periodic orbit rather than on the attractor.

13

Dimensions

Very often a strange attractor is a fractal object, whose geometric structure is invariant under the time evolution maps. In chapter 4 we introduced some concepts related to these fractal properties, and we have also mentioned that the *information dimension* $\dim_H \rho$ of a measure ρ is defined by

$$\dim_H \rho = \inf\{\dim_H S : \rho(S) = 1\}, \tag{92}$$

where $\dim_H S$ is the Hausdorff dimension of the set S (see chapter 4). An interesting fact is the following: let $\rho[B_x(r)]$ be the mass of the measure ρ contained in a ball of radius r centered at x. Then the next proposition gives a characterisation of $\dim_H \rho$ which is very practical from the point of view of numerical computations.

Theorem (Young, 1982)

Let ρ be an ergodic probability measure on a finite-dimensional manifold M, and suppose that

$$\lim_{r \to 0} \frac{\log \rho[B_x(r)]}{\log r} = \alpha \tag{93}$$

for ρ-almost all x. Then $\dim_H \rho = \alpha$. Since ρ is ergodic the existence of the limit (93) implies that it is constant, defining an unique dimension α, called the *information dimension* of ρ.

In an experimental situation one deals with a time series $(x_i)_{1 \leq i \leq N}$ corresponding to measurements regularly spaced in time, and estimates $\rho[B_x(r)]$ by

$$\frac{1}{N} \sum_{j=1}^{N} \theta(r - |x_j - x_i|),$$

where θ is the Heaviside function. Another method, due to Grassberger and Procaccia (1983a,b), is the following. Define

$$C(r) = \frac{1}{N^2} \sum_{i,j} \theta(r - |x_j - x_i|) \quad (N \text{ large}) \tag{94}$$

then

$$\text{correlation dimension} = \lim_{r \to 0} \frac{\log C(r)}{\log r}. \tag{95}$$

Note that the correlation dimension might be different from the information dimension. Usually one computes $C(r)$ using a time series that lives in R^{d_E}, whose points are obtained from the signal (x_i) by the time delay method. Then one plots $\log C(r)$ as a function of $\log r$, for different values of the embedding dimension d_E.

For small r there is a scatter of points due to poor statistics. Increasing r, there is a range (r_0, r_1) where the slope is nearly constant, representing the correlation dimension (if d_E is sufficiently large). This is the 'meaningful' range. For r larger than r_1 there is a deviation from constancy due to nonlinear effects (see Fig. 29).

13.1 Strange attractors as inhomogeneous fractals

Let us come back to Young's theorem. In the case considered there the mass $\rho[B_x(r)]$ satisfies a scaling law of the form

$$\rho[B_x(r)] \sim r^\alpha \tag{96}$$

for ρ-almost all x. When (96) holds with the same α for all points of an attractor, we say that this attractor is a *homogeneous fractal*. However, it may happen, and actually it happens very often, that the attractor is an *inhomogeneous fractal* (or *multifractal*) and that

$$\rho[B_x(r)] \sim r^{\alpha(x)}, \tag{97}$$

namely that the scaling index α fluctuates depending on the particular ball we consider. The fluctuations of α can be characterised by the 'anomalous' scaling laws of the moments of the mass:

$$\langle \rho[B_x(r)]^q \rangle \sim r^{\phi(q)}. \tag{98}$$

It is thus possible to define a hierarchy of *Renyi dimensions* simply by (see Grassberger, 1983; Paladin and Vulpiani, 1987)

$$D_{q+1} = \frac{\phi(q)}{q}. \tag{99}$$

Fig. 29. Experimental result from Malraison *et al.* (1983) and Atten *et al.* (1984). (a) The plots show log C versus log r for different values of the embedding dimension (d_E) for the Rayleigh–Bénard experiment. (b) The measured dimension α as a function of the embedding dimension d_E, both for the Rayleigh–Bénard experiment and for numerical white noise. In the latter case α is nearly equal to d_E. Symbols: + white noise; ●Rayleigh–Bénard.

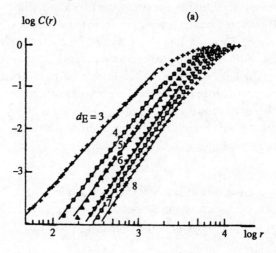

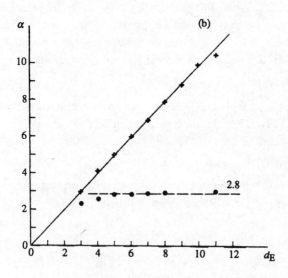

A heuristic argument then shows that the Kolmogorov capacity, the information dimension and the correlation dimension are, respectively, the Renyi dimensions D_0, D_1 and D_2. Notice that for a homogeneous fractal, such that (96) holds uniformly for all points, all the Renyi dimensions are equal to the information dimension.

The multifractal nature of a strange attractor A is related to a spectrum of values of α which is associated with the invariant measure ρ carried by A. We can thus consider (Benzi *et al.*, 1984; Badii and Politi, 1987; Grassberger *et al.*, 1987) a partition Σ of the attractor into pieces Σ_α where the probability measure ρ scales with an exponent α in the interval $\{\alpha, \alpha + d\alpha\}$:

$$\Sigma_\alpha = \{x \in A : \rho[B_x(r)] \sim r^\alpha \text{ for } r \to 0\}. \tag{100}$$

Some authors (Halsey *et al.*, 1986) have proposed to look at the Hausdorff dimension of the sets Σ_α as a function $f(\alpha)$ of α, called the *dimension spectrum*. The interesting fact is that one can deal with this problem by a statistical mechanics procedure, namely by defining a partition function $Z^N(\beta)$ defined for $\beta \in R$ by

$$Z^N(\beta) = \sum \rho(A_i)^\beta, \tag{101}$$

the sum being taken over all sets A_i of a partition of the attractor into sets of size 2^{-N}. The free energy of ρ for this partition $\{A_i\}$ is defined (when it exists) by

$$F(\beta) = \lim_{N \to \infty} \frac{1}{N} \log_2 Z^N(\beta). \tag{102}$$

Then the analogy with statistical mechanics is used to relate the Legendre transform of F (i.e. $\inf_\beta[\beta t - F(\beta)]$) to the dimension spectrum $f(\alpha)$. Recently, this approach has been made rigorous for the case of expanding Markov maps (see Collet, Lebowitz and Porzio, 1987).

13.2 Information dimension, entropy and characteristic exponents

We show now that a natural relation exists between information dimension and the quantities we have previously introduced: entropy and characteristic exponents (see, for instance, Young, 1982, 1984). If ρ is a probability measure carried by an attractor, and

$\lambda_1, \ldots, \lambda_r$ are the characteristic exponents of ρ, we define the *Liapunov dimension* to be

$$\dim_\Lambda \rho = k + \frac{\lambda_1 + \cdots + \lambda_k}{|\lambda_{k+1}|}, \tag{103}$$

where $k = \max\{i: \lambda_1 + \cdots + \lambda_i > 0\}$ (see Fig. 30).

Then Kaplan, Yorke and others have made the following guess, which has led to a lot of important work:

> **Conjecture (Kaplan and Yorke, 1979; Frederickson et al., 1983; Alexander and Yorke, 1984)**
> If ρ is an SRB measure, then 'in general'

$$\dim_H \rho = \dim_\Lambda \rho. \tag{104}$$

This equality holds in a number of situations, but there are also exceptions.

Let us consider now the *partial dimensions* $D^{(i)}$, that can be roughly defined as the Hausdorff dimensions in the directions of the various λ_i

Fig. 30. Determination of the Liapunov dimension. c_ρ is defined by $c_\rho(s) = \Sigma_{i=1} \lambda_i + (s-k)\lambda_{k+1}$ if $k \leqslant s \leqslant k+1$. The number of positive characteristic exponents (unstable dimension) is m_+. The graph is from Manneville (1985, for the Kuramoto–Sivashinsky model.

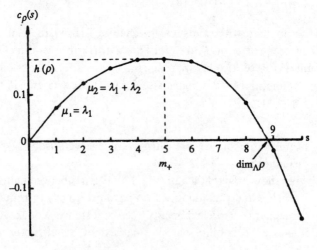

(for more precise definitions see the original paper of Ledrappier and Young, 1985, or Eckmann and Ruelle, 1985).

The $D^{(i)}$ satisfy $0 \leq D^{(i)} \leq m^{(i)}, i = 1, \ldots, r$, where $m^{(i)}$ is the multiplicity of $\lambda^{(i)}$; and the entropy can be expressed in terms of the $D^{(i)}$'s by

$$h(\rho) = \sum_+ \lambda^{(i)} D^{(i)} = -\sum_- \lambda^{(i)} D^{(i)}, \tag{105}$$

where Σ_+ ((Σ_-)) is the sum over positive (negative) characteristic exponents. In particular,

$$\sum_i \lambda^{(i)} D^{(i)} = 0. \tag{106}$$

The relation with information dimension is

$$\dim_H \rho \leq \sum_i D^{(i)}. \tag{107}$$

Then, from the analysis of these partial dimensions follows that, if the Kaplan–Yorke conjecture holds for a measure ρ, then we have $D^{(i)} = m^{(i)}$ for $i = 1, \ldots, k$ and $D^{(i)} = 0$ for $i = k+2, \ldots, r$, the measure ρ is an SRB measure and, as a consequence, the Pesin identity (89) holds.

More generally we have the following inequality:

Theorem (Ledrappier, 1981)

Let f be a C^2 map and ρ an ergodic measure with compact support. Then

$$\dim_H \rho \leq \dim_\Lambda \rho. \tag{108}$$

This inequality, as well as Ruelle's inequality (88), are a natural consequence of the existence of the partial dimensions $D^{(i)}$.

An equality holds in some special cases:

Theorem (Young, 1982)

Let f be a C^2 diffeomorphism of a compact *surface* M and ρ an ergodic measure with compact support and with characteristic exponents $\lambda_1 > 0 > \lambda_2$. Then

$$\dim_H \rho = h(\rho) \left(\frac{1}{\lambda_1} + \frac{1}{|\lambda_2|} \right). \tag{109}$$

In particular, if ρ is an SRB measure, hence $h(\rho) = \lambda_1$, then $\dim_H \rho = \dim_\Lambda \rho$. A formula similar to (109) holds in the following example.

Example 5

Given $x \in [0,1]$, let $x = 0.\omega_1 \omega_2 \ldots$ be the dyadic expansion of x and X_p the set of points in $[0,1]$ that contains 1 in proportion p in their dyadic expansion,

$$X_p = \{x \in [0,1]: \lim_{n \to \infty} \frac{1}{n} \sum_{i=0}^{n-1} \omega_i = p\}. \tag{110}$$

In 1949 Eggleston proved that

$$\dim_H X_p = \frac{1}{\log 2}[-p \log p - (1-p)\log(1-p)]. \tag{111}$$

Notice that $\dim_H X_{\frac{1}{2}} = 1$. Let us consider now the map $f(x) = 2x \pmod 1$ of the unit interval into itself. We know from Example 1 that this map operates as a Bernoulli shift on the dyadic expansion of the points x in $[0,1]$. Let ρ be the unique invariant probability measure on $[0,1]$ satisfying

$$\rho\{x \in [0,1]: \omega_1 = 1\} = p.$$

(For $p = \frac{1}{2}$, ρ corresponds to the Lebesgue measure.) The characteristic exponent of ρ is $\lambda = \log(df/dx) = \log 2$ and the entropy $h(\rho)$ is given by

$$h(\rho) = -\sum_{i=1}^{2} p_i \log p_i,$$

where $p_1 = p$ and $p_2 = 1 - p$. Then, the result of Eggleston can be rephrased, saying that there is a set $X \in [0,1]$ for which $\rho(X) = 1$ and

$$\dim_H(X) = \frac{h(\rho)}{\lambda}. \tag{112}$$

14

Resonances

To conclude we give a brief discussion of another interesting indicator for the statistical analysis of time evolutions: the complex singularities of the power spectrum of a signal. Poles at complex values of the frequency are naturally interpreted as *resonances* of the dynamical system and they are strictly related to the decay properties of the time correlation functions, namely to the *mixing* properties of the time evolution.

Let f be a transformation preserving the ergodic probability measure ρ on a compact manifold M. We call *observable* any differentiable function on M. Then, given two observables B, C, we introduce the *correlation function* $C_{BC}(t)$ by

$$C_{BC}(t) = \rho[(B \circ f^t)C] - \rho(B)\rho(C). \tag{113}$$

Its Fourier transform is

$$S_{BC}(\omega) = \int_{-\infty}^{+\infty} dt \exp(i\omega t) C_{BC}(t). \tag{114}$$

Notice that $S_{BB}(\omega)$ is the power spectrum of the signal $B(x(t))$. Some theoretical work has been carried out recently for Axiom A systems (see Pollicot, 1976; Ruelle, 1985, 1986) through a statistical mechanics formalism where the *time* correlation functions of the dynamical system (Axiom A) are analysed in terms of the *space* correlations of a one-dimensional lattice system with short range interactions. In this framework the authors proved that the Fourier transform of the correlation function is meromorphic (i.e. holomorphic except for poles) in a strip, and the position of the poles does not depend on the observables monitored. Moreover, for a discrete-time mixing system (i.e. a system for which $C_{BC}(t) \to 0$ when $|t| \to \infty$ for all B,C) there are no poles in a sufficiently narrow strip above and below the real axis, and this corresponds to an exponential decay of correlations.

For more general (non-Axiom A) dynamical systems, as well as for experimental time series, numerical studies are often the only accessible tool of investigation. Consider a time series $(x_n)_{1 \leqslant n \leqslant N}$, the correlation function is given by

$$C(t,T) = (\text{const.})[\langle x(\tau + t)x(\tau) \rangle - \langle x(\tau) \rangle^2], \tag{115}$$

where $x(\tau) \equiv x(n\Delta\tau) = x_n$ ($\Delta\tau$ is the time spacing), $T = N\Delta\tau$. The brackets indicate the time average along the time evolution (up to time T). For a chaotic system the largest characteristic exponent λ_1 is positive and a small imprecision ε on x will grow like $\varepsilon \exp(\lambda_1\tau)$. Thus, if m denotes the maximum correlation time t, and ε the noise level due to round off errors (or to experimental noise), a precise determination of $C(t) = C(t,T \to \infty)$ requires

$$m \ll \frac{|\log \varepsilon|}{\lambda_1} \ll T. \tag{116}$$

The second inequality ensures the selection of the physical measure ρ (according to the discussion in chapter 5) and the first one ensures that one observes only the intrinsic decay properties of $C(t)$.

As we have seen in chapter 1, the spectrum of a chaotic motion is usually continuous in ω and we can write (in the limit $T \to \infty$) for a discrete-time dynamical system:

$$C(t) = \frac{1}{2\pi} \int_0^{2\pi} S(\omega)\exp(i\omega t), \tag{117}$$

where

$$S(\omega) = \sum_{t=-\infty}^{\infty} C(t)\exp(-i\omega t). \tag{118}$$

It is clear from (117) that the asymptotic behavior of the correlation function is determined by the singularities of the analytic continuation of $S(\omega)$, which are nearest to the real axis: if the nearest singularity is a simple pole located at $z_0 = x_0 + iy_0$, then, for large t, one gets

$$C(t) \sim \exp(-y_0 t)\cos(x_0 t + \phi). \tag{119}$$

So, while x_0 determines the frequency of the leading resonance, y_0 turns out to drive the exponential damping of the corresponding oscillation of $C(t)$. As a matter of fact, some recent numerical studies (Isola, 1987), using the technique of Padé approximants, have shown the presence of resonances in the Hénon system emphasising their role in the determination of the type of mixing rate.

15

Conclusions

This review has led from those general ideas which constitute the major breakthroughs in the physical theory of dynamical systems – deterministic noise, low-dimensional chaos, strange attractors and so on – to a discussion of those quantities accessible at present through the *statistical analysis of time series* for deterministic nonlinear systems. The discussion presented here is fairly incomplete: some other indicators have been recently introduced and utilised for the study of chaotic time series (on the other hand the work on this matter is rapidly evolving and therefore it is quite hopeless to give an exhaustive picture of the situation). Moreover we have somewhat neglected the discussion of the analysis of actual experimental data. At this time one of the most promising directions of research is in the systematic application of the ideas and techniques of nonlinear dynamics to all kinds of natural systems, including those of biology, the social sciences and economics (for the latter see, for instance, Scheinkman and Le Baron, 1987). Many examples of aperiodic behavior can be tested for the presence of low-dimensional deterministic chaos, and negative as well as positive answers will be informative.

REFERENCES

Alexander, J.C. and Yorke, J.A. 1984, *Ergod. Th. Dyn. Syst.* **4**, 1.

Anderson, P.W. 1958, *Phys. Rev.* **109**, 1492.

Arnold, V. 1980, *Chapitres Supplémentaires de la Théorie des Équations Différentielles Ordinaires*, Editions Mir, Moscow, P. 129.

Atten, P., Caputo, J.C., Malraison, B. and Gagne, Y. 1984, preprint.

Badii, R. and Politi, A. 1987, *Physica Scripta*, **35**, 243.

Benzi, R., Paladin, G., Parisi, G. and Vulpiani, A. 1984, *J. Phys.* **A17**, 3521.

Billingsley, P. 1965, *Ergodic Theory and Information*, Wiley, New York.

Bowen, R. 1975, *Equilibrium States and the Ergodic Theory of Ansov Diffeomorphisms*, Lecture Notes in Mathematics 470, Springer, Berlin.

Bowen, R. 1978, *On Axiom A Diffeomorphisms*, Regional Conference Series in Mathematics, no. 35, American Mathematics Society, RI.

Bowen, R. and Ruelle, D. 1975, *Inv. Math.* **29**, 181.

Caffarelli, R., Kohn, R. and Nirenberg, L. 1982, *Comm. Pure. Appl. Math.* **35**, 771.

Collet, P. and Eckmann, J.-P. 1980a, *Comm. Math. Phys.* **76**, 115.

Collet, P. and Eckmann, J.-P. 1980b, *Iterated Maps on the Interval as Dynamical Systems*, Birkhauser, Cambridge, MA.

Collet, P. and Eckmann, J.-P. 1983, *Ergod. Th. Dyn. Syst.* **3**, 13.

Collet, P., Lebowitz, J.L. and Porzio, A. 1987, *J. Stat. Phys.* **47**, 609.

Curry, J.H. 1979, *Comm. Math. Phys.* **68**, 129.

Eckmann, J.-P. 1981, *Rev. Mod. Phys.* **53**, 643.

Eckmann, J.-P., Oliffson-Kamphorst, S., Ruelle, D. and Ciliberto, S. 1986, *Phys. Rev.* **A34**, 4971.

Eckmann, J.-P. and Ruelle, D. 1985, *Rev. Mod. Phys.* **57**, 617.

Feit, S.D. 1978, *Comm. Math. Phys.* 61, 249.

Frederickson, P., Kaplan, J.L., Yorke, E.D. and Yorke, J.A. 1983, *J. Diff. Eq.* **49**, 185.

Furstenberg, H. 1963, *Trans. Am. Math. Soc.* **108**, 377.

Gollub, J.P. and Swinney, H.L. 1978, *Phys. Today* **31**, 41.

Grassberger, P. 1983, *Phys. Lett.* **97A**, 227.

Grassberger, P. 1984, preprint, Wuppertal.

Grassberger, P. 1986, preprint, Wuppertal.

Grassberger, P., Badii, R. and Politi, A. 1987, preprint, Wuppertal.

Grassberger, P. and Procaccia, I. 1983a, *Phys. Rev.* **A28**, 2591.

Grassberger, P. and Procaccia, I. 1983b, *Physica* **D9**, 189.

Guckenheimer, J. and Holmes, P. 1983, *Nonlinear Oscillations, Dynamical Systems and Bifurcations of Vector Fields*, Springer, Berlin.

Halsey, T.C., Jensen, M.H., Kadanoff, L.P., Procaccia, I. and Shraiman, B. 1986, *Phys. Rev.* **A33**, 1141.

Hénon, M. 1976, *Comm. Math. Phys.* **50**, 69.

Isola, S. 1988, *Comm. Math. Phys.* (in press).

Jakobson, M. 1981, *Comm. Math. Phys.* **81**, 39.

Janich, J. 1984, *Topology*, Undergraduate Texts in Mathematics, Springer, Berlin.

Kaplan, J.L. and Yorke, J.A. 1979, *Comm. Math. Phys.* **67**, 93.

Kifer, Y.I. 1974, preprint.

Kotani, S. 1987, *Contemp. Math.* **50**, 277.

Lanford, O.E. 1977, *Lecture Notes Math.* **615**, 113.

Ledrappier, F. 1981, *Comm. Math. Phys.* **81**, 229.

Ledrappier, F. and Young. L.-S. 1985, *Ann. Math.* **122**, 509 and 540.

Lorenz, E.N. 1963, *J. Atmos. Sci.* **20**, 130.

Malraison, B., Atten, P., Bergé, P. and Dubois, M. 1983, *J. Phys. Lett.* **44**, 897.

Mandelbrot, B. 1982, *The Fractal Geometry of Nature*, Freeman, San Francisco.

Mañé, R. 1981, *Lecture Notes Math.* **898**, 230.

Manneville, P. 1985, preprint.

May, R.M. 1976, *Nature*, **261**, 459.

Oseledec, V.I. 1968, *Trudy Mosk, Mat. Obšč.* **19**, 179.

Packard, N.H., Crutchfield, J.P., Farmer, J.D. and Shaw, R.S. 1980, *Phys. Rev. Lett.* **45**, 712.

Paladin, G. and Vulpiani, A. 1987, *Phys. Rep.* **156**, (4), 149.

Pesin, Y.B. 1977, *Russ. Math. Surv.* **32**, (4), 55.

Pollicott, M. 1976, *Inventh. Math.* **34**, 231.

Pugh, C.C. and Shub, M. 1984, preprint.

Roux, J.-C. and Swinney, H.L. 1981, in *Nonlinear Phenomena in Chemical Dynamics*, ed. C. Vidal and A. Pacault, Springer, Berlin, p. 38.

Ruelle, D. 1976, *Am. J. Math.* **98**, 619.

Ruelle, D. 1978, *Bol. Soc. Bras. Mat.* **9**, 83.

Ruelle, D. 1981, *Comm. Math. Phys.* **82**, 137.

Ruelle, D. 1985, *Phys. Rev. Lett.* **56**, 405.

Ruelle, D. 1986, *J. Stat. Phys.* **44**, 281.

Ruelle, D. and Takens, F. 1971, *Comm. Math. Phys.* **21**, 12.

Shaw, R.S. 1981, *Z. Naturforsch.* **36a**, 80.

Scheinkman, J.A. and Le Baron, B. 1987, preprint, Chicago.

Shaw, R.S. in *Chaos and Order in Nature*, ed. H. Haken, Springer, Berlin.

Sinai, J.G. 1972, *Russ. Math. Surv.* **27**, (4), 21.

Smale, S. 1967, *Bull. Am. Math. Soc.* **73**, 747.

Takens, F. 1981, in *Dynamical Systems and Turbulence, Warwick 1980*, ed.
D. Rand and L.-S. Young, Lecture Notes in Mathematics 898,
Springer, Berlin.

Thom, R. 1975, *Structural Stability and Morphogenesis*, Benjamin, New
York.

Thouless, D.J. 1972, *J. Phys. C* **5**, 77.

Ulam, S.M. and von Neumann, J. 1947, *Bull. Am. Math. Soc.* **53**, 1120.

Young, L.-S. 1982, *Ergod. Th. Dyn. Syst.* **2**, 109.

Young, L.-S. 1984, *Physica* **A124**, 639.

BIBLIOGRAPHY

The books suggested here are relevant to the material in this book.

Abraham, R.H. and Shaw, C.D. 1981, *Dynamics – The Geometry of Behavior*, Aerial Press, Santa Cruz.

Arnold, V. 1980, *Chapitres Supplémentaires de la Théorie des Equations Différentielles Ordinaires*, Editions Mir, Moscow, p. 129.

Arnold, V.I. and Avez, A. 1968, *Ergodic Problems of Classical Mechanics*, Benjamin, New York.

Bergé, P., Pomeau, Y. and Vidal, C. 1984, *L'ordre dans le Chaos*, Hermann, Paris.

Billingsley, P. 1965, *Ergodic Theory and Information*, Wiley, New York.

Bowen, R. 1975, *Equilibrium States and the Ergodic Theory of Ansov Diffeomorphisms*, Lecture Notes in Mathematics 470, Springer, Berlin.

Bowen, R. 1978, *On Axiom A Diffeomorphisms*, Regional Conference Series in Mathematics, no. 35, American Mathematics Society, RI.

Collet, P. and Eckmann, J.-P. 1980, *Iterated Maps on the Interval as Dynamical Systems*, Birkhauser, Cambridge, MA.

Cvitanović, P., ed. 1984, *Universality in Chaos*, Adam Higler, Bristol.

Guckenheimer, J. and Holmes, P. 1983, *Nonlinear Oscillations, Dynamical Systems and Bifurcations of Vector Fields*, Springer, Berlin.

Hao Bai-Lin, 1984, *Chaos*, World Scientific, Singapore.

Janich, J. 1984, *Topology*, Undergraduate Texts in Mathematics, Springer, Berlin.

Les Houches, 1983, *Chaotic Behaviour of Deterministic Systems*, Session XXXVI, 1981, North-Holland Publication Corporation.

Mandelbrot, B. 1982, *The Fractal Geometry of Nature*, Freeman, San Francisco.

Mañé, R. 1987, *Ergodic Theory and Differentiable Dynamics*, Springer, New York.

Roux, J.-C. and Swinney, H.L. 1981, in *Nonlinear Phenomena in Chemical Dynamics*, ed. C. Vidal and A. Pacault, Springer, Berlin, p. 38.

Ruelle, D. 1978, *Thermodynamic Formalism*, Encyclopedia of Mathematics and its Applications, vol. 5, Addison-Wesley, Reading, MA. Reissued in 1984 by Cambridge University Press.

Schuster, H.G. 1988, *Deterministic Chaos, an Introduction*, vol. 2, revised edn., VCH, Weinhem.

Shaw, R.S. in *Chaos and Order in Nature*, ed. H. Haken, Springer, Berlin.

Smale, S. 1980, *The Mathematics of Time*, Springer, Berlin.

Takens, F. 1981, in *Dynamical Systems and Turbulence, Warwick 1980*, ed. D. Rand and L.-S. Young, *Lecture Notes in Mathematics 898*, Springer, Berlin.

Thom, R. 1975, *Structural Stability and Morphogenesis*, Benjamin, New York.

Shannon, C.E. and Weaver, W. 1949, *The Mathematical Theory of Communication*, University of Illinois Press.

INDEX